Molecularly Imprinted Polymers as Advanced Drug Delivery Systems

Zhaosheng Liu • Yanping Huang • Yi Yang
Editors

Molecularly Imprinted Polymers as Advanced Drug Delivery Systems

Synthesis, Character and Application

Springer

Editors
Zhaosheng Liu
College of Pharmacy
Tianjin Medical University
Tianjin, China

Yanping Huang
College of Pharmacy
Tianjin Medical University
Tianjin, China

Yi Yang
School Light Industry and Chemical Engineering
Dalian Polytechnic University
Dalian, China

ISBN 978-981-16-0226-9 ISBN 978-981-16-0227-6 (eBook)
https://doi.org/10.1007/978-981-16-0227-6

This Springer imprint is published by the registered company Springer Nature Singapore Pte Ltd.
The registered company address is: 152 Beach Road, #21-01/04 Gateway East, Singapore 189721, Singapore

Preface

The aim of the present book is to bring into focus an attractive research area of molecular imprinting in drug delivery systems. Comprising a diverse group of experts from prestigious universities, the contributors to this book provide access to the latest knowledge and eye-catching achievements in the field and an understanding of what progress has been made and to what extent it is being advanced in current research status. The book provides an in-depth review of the general principles of molecular imprinting technology, preparation process, basic characteristics, and the current research status in drug delivery systems. The use of molecularly imprinted polymers (MIPs) in drug delivery systems allows devising important materials with technical details, including enantioselective MIPs, water-compatible MIPs, and MIPs that specifically respond to specific environmental stimuli. In addition, the use of MIPs for transdermal drug delivery, anticancer preparations and ophthalmic drug delivery, release mechanism, in vivo and in vitro bioavailability evaluation, and future development trends are also discussed. It serves as an important reference for scientists, students, and researchers who are working in the areas of molecular imprinting, drug delivery systems, molecular recognition, materials science, biotechnology, and nanotechnology.

Tianjin, China — Zhaosheng Liu
Tianjin, China — Yanping Huang
Dalian, China — Yi Yang

Contents

1 **Introduction** . . . 1
Dongyu Gu and Yi Yang

2 **Synthetic Strategies for the Generation of Molecularly Imprinted Polymers** . . . 27
Xi Wu

3 **Special Control by Molecularly Imprinted Materials-Zero-Order Sustained Release, Enantioselective MIPs, and Self-Regulated Drug Delivery Microdevices** . . . 61
Xue Zhang, Xiao Liu, Ze-Hui Wei, and Yan-Ping Huang

4 **Water Compatible Molecularly Imprinted Polymers** . . . 77
Qiliang Deng

5 **Stimuli Responsive Imprinted DDS** . . . 93
Li-Ping Zhang and Zhaosheng Liu

6 **MIP as Drug Delivery Systems for Dermal Delivery** . . . 111
Zehui Wei, Lina Mu, and Zhaosheng Liu

7 **MIP as Drug Delivery Systems of Anticancer Agents** . . . 133
Jing Feng and Zhaosheng Liu

8 **MIP as Drug Delivery Systems of Ophthalmic Drugs** . . . 153
Long Zhao and Zhaosheng Liu

9 **MIP as Drug Delivery Systems for Special Application** . . . 179
Li Ma and Zhaosheng Liu

10 **Outlook** . . . 201
Xuemei Wang, Pengfei Huang, and Zheng Zhou

Chapter 1
Introduction

Dongyu Gu and Yi Yang

1.1 A Brief History of Imprinting

MIPs are typical artificial molecular recognition systems, which can be used for specific recognition of target molecules [1–4]. Actually, the subtle molecular recognition, an important basis of life, widely exits in nature [5]. The discovery process of molecular recognition originated from the early 1890s. von Behring and Kitasato observed serum transferred from immunized rabbits could cure diptheria and used the term "antitoxin" to refer to a mediator in the body neutralized the toxins [6]. Subsequently, Ehrlich used the term "antibody" to replace "antitoxin" [7] and proposed side-chain theory for antibody-antigen binding in a "lock-and-key" interaction [8]. The concept of molecular imprinting was first proposed using silica matrices by Polyakov in 1931. Silica particles were prepared from sodium silicate and took benzene, toluene, or xylene as organic additives. The results showed that the uptake capacity for the associated additive increased compared with the other two structural analogues [9]. In 1940, Pauling proposed the theory of the structure and process of formation of antibodies. The denaturing reagents were added into the medium to cause the unfolding of protein in the presence of a foreign antigen. Then, the protein molecule refolded to a configuration complementary to the template antigen, thus acquiring the homologous antibody [10]. Subsequently, this procedure was demonstrated again through the production of antibody in vitro using γ-globulin [11]. Although Pauling's theory was denied by Burnet's clonal selection theory [12], these studies about the production of antibodies can be considered as the classic

D. Gu
College of Marine Science and Environment, Dalian Ocean University, Dalian, China

Y. Yang (✉)
School of Light Industry and Chemical Engineering, Dalian Polytechnic University, Dalian, China
e-mail: yangyi105@mails.ucas.ac.cn

Z. Liu et al. (eds.), *Molecularly Imprinted Polymers as Advanced Drug Delivery Systems*, https://doi.org/10.1007/978-981-16-0227-6_1

cases of early bio-imprinting [3] to inspire scientists to translate them into practical applications.

However, the limited knowledge became the main barrier for developing the new technology [3]. Until 1949, after communicating with Pauling, Dickey prepared the adsorbents with the automatically forming pockets using the initial mixture of sodium silicate and glacial acetic acid in the presence of alkyl orange dyes. The pockets in the adsorbent had the specific affinities with the dyes and foreign molecular which demonstrated the existence of imprinting effect. Moreover, the results indicated that the alkyl groups on the tertiary amine had an important influence on the imprinting effect [13]. It is no doubt that this result is very important, but no relevant research was reported in the next 2 years. In 1952, despite the lower enrichment ratio, the imprinted silica was applied to enrich the stereoisomers of camphorsulfonic and mandelic acids for the first time [14]. In 1955, Haldeman and Emmett mentioned the word "imprints" for first time in their study [15]. Although several applications of imprinted silica materials were investigated in the following years [16, 17], due to the instability of the imprinted cavities and the limited number of feasible templates, there is no major breakthrough in this field [17].

In 1972, Wulff's group prepared covalent molecular imprinting polymers (MIPs) successfully for the first time, which has made a great breakthrough in the field of imprinting technology [3, 18]. In the preparation of these covalent imprinting polymers, functional monomer and template were connected by covalent linkage which must be stable enough to keep intact during the polymerization, but had to be cleaved easily at the same time [19]. In the following years, more similar studies further promoted the development of covalent molecular imprinting technology (MIT) [20, 21]. In their study, the covalent macroporous polymers with chiral cavities were prepared using 4-nitrophenyl-α-D-mannopyranosid as template for the separation of the enantiomers. The optical enrichment in a chromatographic separation amounted to 87% [22]. However, the disadvantage of this method was that the guest binding and release were too slow [19]. As a result, progress in this period was relatively slow.

A decade later, the non-covalent imprinting molecules were developed in the 1880s. As long as some kind of non-covalent interactions between functional monomer and template exist, such as hydrogen bond, hydrophobic force, van der Waals force, etc. the imprinting molecular can be formed. Furthermore, the template can be easily removed by simple extraction after the polymerization [23, 24]. Since then, this technology had attracted much attention. Non-covalent imprinting created the tailor-shaped cavities with high specificity and affinity for a target molecule, which was successfully applied for the separation of racemates, the understanding of enzymatic action and the use of imprinted system as catalysts [25, 26]. Subsequently, the same strategy was used to obtain the materials of various formats, which expanded the application scope in different important fields [27]. Though the preparative procedures for non-covalent MIPs were far simpler than those for covalent imprinting molecules [19], a more uniform rebinding cavities and more target-specific MIPs can be produced by the covalent synthesis [4].

In 1995, a new approach hybridized covalent imprinting and non-covalent imprinting was developed to introduce the recognition site functionality into polymers, which combined the advantages of clear-cut nature in the covalent imprinting and fast guest binding in the non-covalent imprinting. The polymers displayed characteristics similar to a true biological receptor or synthetic host for cholesterol [28]. Nowadays, due to high selectivity, specificity, stability and durability, molecular imprinting has become a hot topic in the study of supramolecular structure in the twenty-first century [19, 29, 30].

1.2 General Principle of Molecular Imprinting

Molecular imprinting technology has evolved a great deal of extensive application to the synthesis of organic and hybrid MIPs with the characteristic of antibodies or enzymes [31, 32]. But all of the methods for preparation of MIPs follow the similar basic outline (Fig. 1.1), including the processes of recognition site formation and template removal [33]. The cross-linker, functional monomer and porogenic solvents polymerized around the template covalently or non-covalently to form the highly cross-linked polymers MIPs [34, 35]. In this way, this material can possess a permanent pore structure and a high internal surface area [36]. The template molecule combines with a complementary functional monomer to produce a constellation of template–monomer complexes (Fig. 1.1). The templates include ion, atom, molecule, complex, macromolecular and microorganisms, etc. [37].

MIPs can be produced by covalent or non covalent combination of template and functional monomer. Covalent imprinting, a typical method, ensures that functional monomer residues exist only in the imprinted cavities. The Ketals/acetals, Schiff's base and boronate esters are often used in the covalent imprinting [22, 38, 39]. But this method is not flexible because the reversible condensation reactions are few. In addition, the slow dissociation and binding caused by strong covalent action makes it difficult to achieve thermal equilibrium [37]. The non-covalent interactions between template and polymerizable functionality include reversible covalent bond(s), covalently attached polymerizable binding groups that are activated for non-covalent interaction by template cleavage, electrostatic, hydrophobic or van der Waals interactions, H-bond(s), p–p interactions or coordination with a metal center, etc. [37]. Hydrogen bonding is the most common and major interaction. For example, it appears in the interaction of methacrylic acid groups and primary amines in nonpolar solvents [40]. The advantage of this method is that there is no kinetic barrier to imprint–functional monomer complex formation and target molecule recognition. On the contrary, the main limiting factor is diffusion, which can be easily mitigated by careful selection of system parameters. Since the interaction in non-covalent imprinting is relatively weak, the excess of functional monomer is required to improve the formation of imprint–functional monomer complex. In recent years, non-covalent imprinting has become the most popular synthetic strategy due to its simple operation, fast binding and removal. But non-covalent MIPs

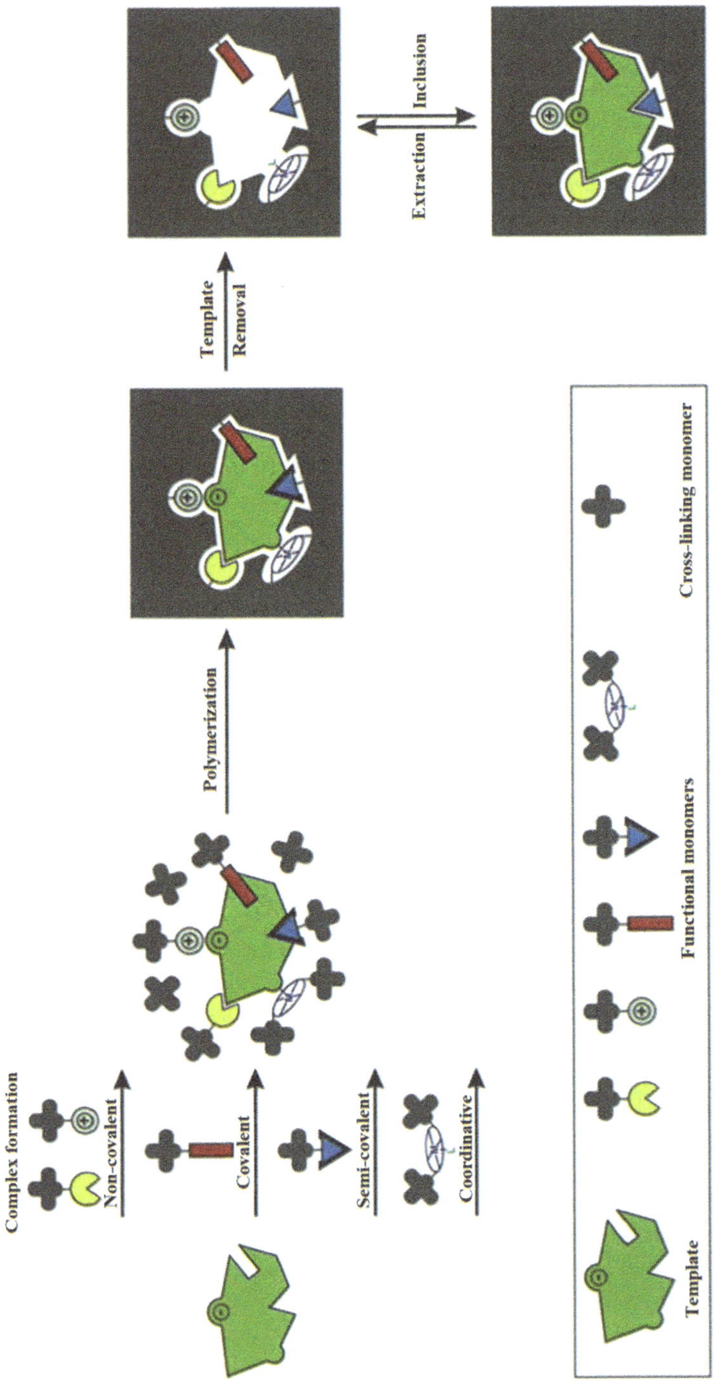

Fig. 1.1 Different routes of molecular cavity formation in a molecularly imprinted polymer. (Reproduced with permission of Ref. [33])

have the disadvantage of instability [41]. So, a new semicovalent imprinting method was developed which combined the durability of covalent imprinting and the rapid target uptake of non-covalent imprinting, a small sacrificial spacer fragment such as carbon dioxide was often used in this method [28]. This method provided an intermediate choice, namely, the template bonded to the functional monomer covalently, but the recombination of template was based on non-covalent interaction. In addition, coordination chemistry can be used to prepare MIPs. Metal ion can form a part of complex that was bound to an imprint cavity covalently and participated in target recognition through metal–ligand bonding interactions, or acted as an ionic template to create an imprint cavity that can interact with an appropriate target metal ion. This method was easy to tailor according to the specific requirements because the selection of the metal and its ligands was wide.

The removal of template is also an important step to vacate target-specific cavity with the geometry and position of the functional groups which can interact with the target molecule. Thus, MIPs can be used for selectively rebinding the template or analogues to achieve imprinting. The target molecule may be the same as the template, but not always the same. The template molecules can be separated efficiently under suitable conditions [36]. The MIP is exposed to the sample containing the target, and the cavity fishes the target molecule from the complex sample selectively. In contrast to biological systems, in which the target must match the antibody or an antibody must be generated specifically for that target, MIPs can be used in almost any target molecule [4].

In the process of MIPs preparation, template and functional monomer can interact through covalent interaction of pre-organized approach, non-covalent interaction of self-assembly approach, hybrid of covalent and non-covalent interactions, and chelation of metal ion. The strength of these interactions depends on the functional groups provided by the functional monomers, which determines the affinity of MIP and the selectivity, and accuracy of recognition site. All kinds of polymerization, including radical, anion, cation and condensation, can be used for molecular imprinting. The common functional monomers were shown as Fig. 1.2 [42].

1.3 The Imprinting Matrix

Both of organic and inorganic materials can be used as matrices for molecular imprinting without special restrictions.

1.3.1 Organic Matrix

The abundance of organic and inorganic polymer precursors leads to a large number of molecular imprinting studies. Although the number of combinations that can produce a MIP is essentially infinite, the vast majority of reports deal with organic

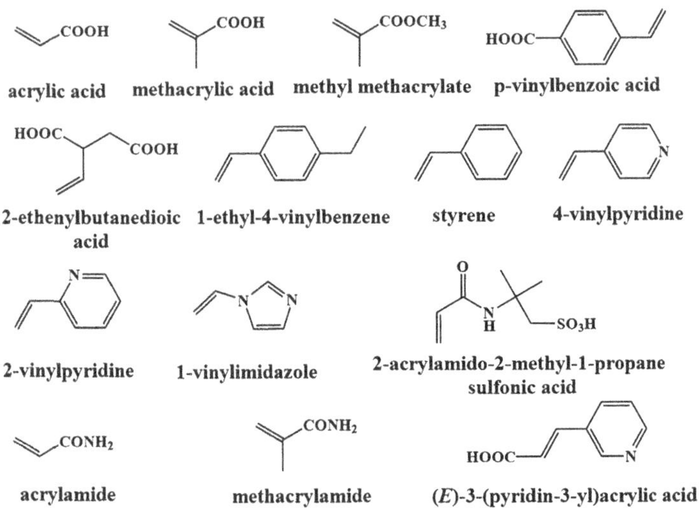

Fig. 1.2 Common functional monomers in molecular imprinting. (Reproduced with permission of Ref. [42])

polymer systems to form a network or matrix. A high degree of crosslinking is required in this procedure to maintain the physical stability of the imprinted cavities for obtaining the MIPs with high specificity [43]. However, the degree of crosslinking is affected by the type and amount of cross-linker. When the ratio of cross-linker to functional monomer is less than 10%, the imprinted polymer and non-imprinted polymer cannot be distinguished, while the selectivity of the MIP is markedly enhanced when the cross-linker ratio is above 50%. The fundamental role of these reagents is to fix the guest-binding sites firmly in the desired structure. In addition, the suitable cross-linker also can control the morphology of the polymer matrix to produce gel-type, macroporous or a microgel powder [43]. There are more than 4000 polymerisable compounds commercially used in the preparation of MIPs. Among them, the major cross-linkers used for MIT were listed in Fig. 1.3 [42].

1.3.2 Inorganic Matrix

Silicon and titanium alkoxides are the most commonly and widely used inorganic gel matrices for MIPs, which can be considered a small niche in MIT studies [44, 45].

1.3.2.1 Silica Gel Matrices

Sol–gel matrix originated from methyl orange as template for preparing MIPs. Compared with non-imprinted gel, more adsorption of template methyl orange

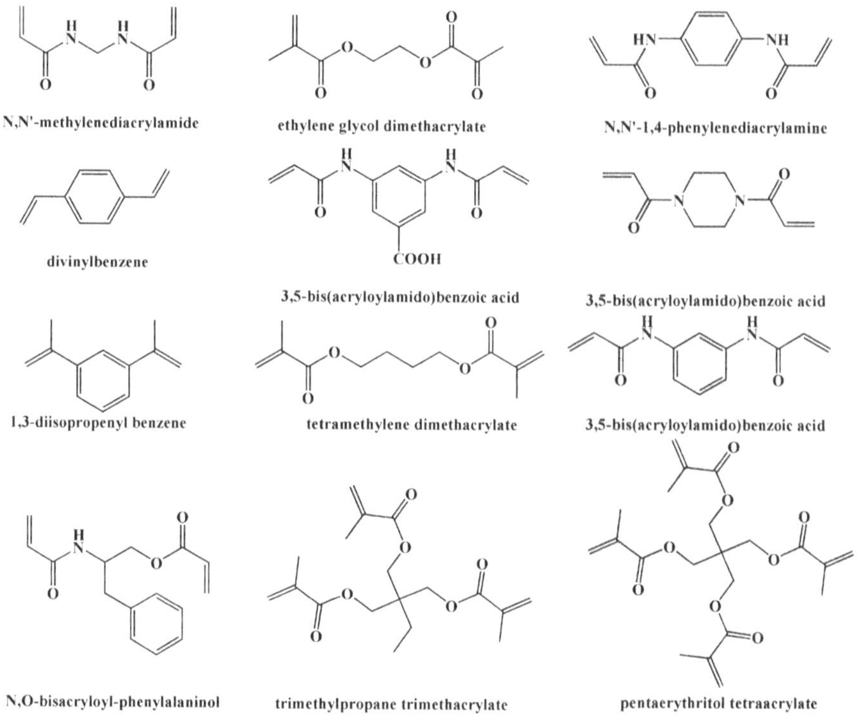

Fig. 1.3 Chemical structures of common cross-linkers for molecular imprinting. (Reproduced with permission of Ref. [42])

was found in imprinted silica gel, which was the first publication of molecular imprinting in sol–gel matrix [13]. The sol–gel methods had created a new way to achieve molecular imprinting in silica. Later, the sol–gel matrices was successfully applied for synthesis of MIPs with (−) menthol, (−)-borneol, (−)-camphor and (+)-fenchol as templates [46, 47]. Compared with Dickey's results, there was no difference in the adsorption of template molecules between the imprinted gel and the non imprinted gel. But, interestingly, the amorphous microporous oxide remembered the kinetic diameter of imprinted molecule (alcohol) [46]. This study provided another way for molecular imprinting to enter inorganic oxides, especially sol–gel materials [48].

In the MIPs synthesis processes, the template was bound to the Si atom in silicon tetraalkoxides, such as tetraethoxysilane (TEOS), tetramethyl-orthosilicate (TMOS), by replacing one of the Si–O bonds with a Si–C bond. The alkoxides function as cross-linkers and the oxide network was formed by polycondensation, while the Si–C bonds remained intact to achieve covalent imprinting [44]. The recognition ability of MIPs was controlled by polymerization conditions, nature of cross-linkers, degree of cross-linking, monomers and nature of template. Due to the high degree of cross-linking, the sol–gel material possessed high thermal stability, and the template can

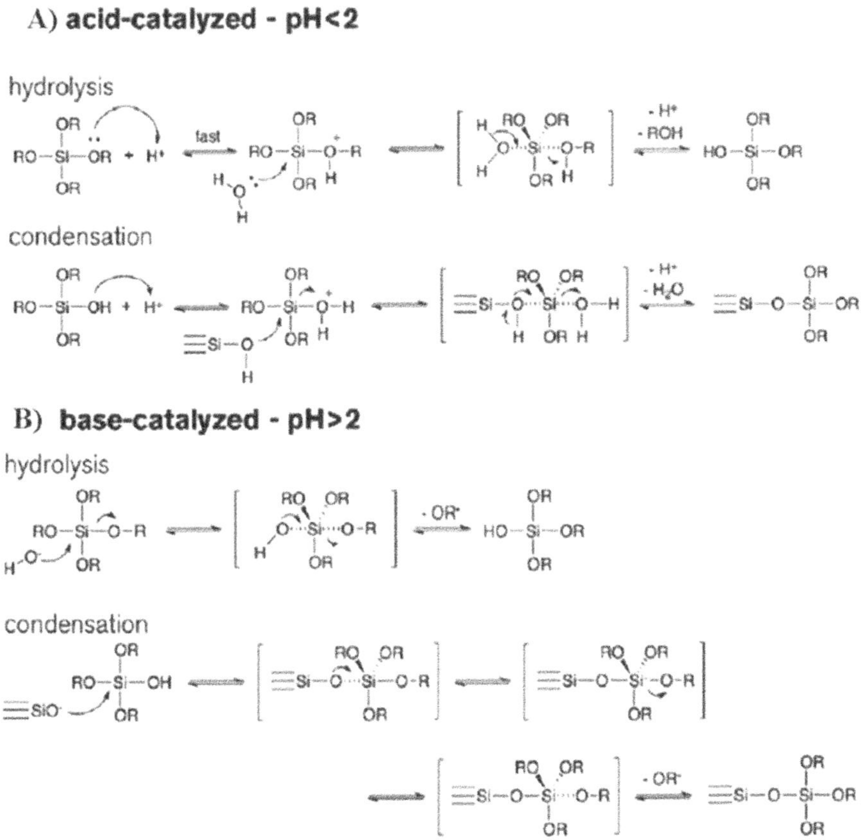

Fig. 1.4 General mechanisms of hydrolysis and condensation of alkoxysilane precursors to form silica in (**a**) acid catalyzed conditions and (**b**) base catalyzed conditions. Condensation can produce either water or alcohol as a byproduct. (Reproduced with permission of Ref. [41])

be removed conveniently using calcinations with high temperature [49]. Porous sol–gel material with an extremely high surface area was very conducive to be an imprinting host [49]. In addition, composition of sol, water to precursor ratio, nature of precursor (polarity), catalyst (acidic or basic), pH, solvent, temperature, humidity and storage conditions could also influence the ability of the MIPs [48, 50]. The reactions were carried out at pH < 2 by acid catalysis or pH > 2 by base catalysis (Fig. 1.4). The isoelectric point of silica was pH $= 2$, and the electric mobility of silica was zero and reaction rates were very low at this pH. In general, the sol–gel process was that the system changed from a liquid "sol" to a solid "gel" phase. In the sol–gel reactions, the water and alcohol were used as porogens in materials, and the material obtained was called a xerogel after drying. If it was dried under supercritical conditions, it will produce an aerogel. Aerogels had very low density, high specific surface area and large porosity. They had excellent thermal insulation and electrical insulation properties. However, their large surface area also made them inferior to xerogels, which had smaller porosity and surface area.

Except for extreme conditions, the reactivity of sol–gel silica was low in all different conditions, including strong acid, strong base, oxidizers, and toxic fluoride species. The highly cross-linked and rigid structure of xerogel silica made it possible to create delicate imprint sites, and had higher shape selectivity compared with more flexible organic polymers. This template fidelity may be the main reason for the success of early silica imprinting. In the sol stage, the shape of silica can be greatly controlled. The minimal swelling property and excellent thermal stability of silica also enabled it to maintain the shape and size of imprint cavities. Silica had few oxidation and ageing problems, which was better than many organic polymers. Silica was also compatible with biological systems and aqueous, and can encapsulate antibodies and enzymes without destroying their activity successfully. In summary, these advantages made sol–gel silica suitable for a wide range of applications [51].

Based on the above works, in the past decade, sol–gel molecular imprinting had made new developments, including organic–inorganic hydrid MIPs [52, 53], surface molecular imprinting technique combined with sol–gel [54], bio-imprinting in sol–gel [55] and molecularly imprinted sol–gel nanotube membrane [56]. The synthesis of organic-inorganic hydrid MIPs (organically modified silanes/ORMOSILS) that combined the properties of inorganic and organic compounds in one material was a very important development in sol–gel process. Due to the variable chemical composition of precursors and the ratio of organic and inorganic start materials, the common inorganic precursors, such as ORMOSILS [52], TMOS, TEOS, tetra-n-propoxysilane and tetrabutoxysilane, had a wide range of properties and applications [48, 50]. In addition, the other metal alkoxides, such as tertrapropyl-orthozirconate (TPOZ), tetraethyl-orthotitanate (TEOT), aluminum *tert*-butoxide, and so forth, can also be used as inorganic precursors [48, 50]. To cope with the undesirable kinetics of the adsorption/desorption and slow mass transfer due to poor accessibility of target molecules which caused by the template and functionality embedded in polymer matrices, surface molecular imprinting was developed. The sol–gel matrices further improved this method to produce 3D matrices in different configurations (bulk structures, thin film and powder) for various sensing application [57]. The surface sol–gel method as a means of molecular imprinting was superior to the previous imprinting technology because the process of gel formation, grinding and extraction took time in the traditional sol–gel process, resulting in the problem of imprinting cavity [45]. Surface molecular imprinting combined with sol–gel can also be used for protein recognition [58]. BSA was imprinted on the functional biopolymer chitosan (CS) microspheres by covalent linkage. These microspheres were surrounded by APTMS and TEOS derived hybrid sol–gel polymeric matrix in aqueous solution at room temperature (Fig. 1.5). The reproducibility and stability of the final material were improved by the grafting of imprinted layer through organic–inorganic hybridization.

The design of molecularly imprinted sol–gel nanotube membranes offered some interesting advantages for biochemical applications. Silica nanotubes with the cross-linked structure were easy to prepare and were highly suitable for the formation of a

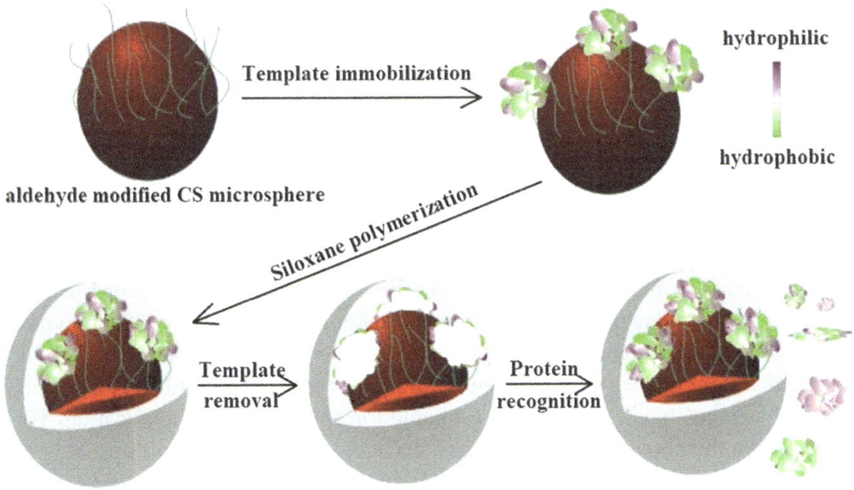

Fig. 1.5 Schematic representation for synthesis of the protein-imprinted polymer on chitosan microsphere using immobilized protein template. (Reproduced with permission of Ref. [58])

delicate recognition site [56, 59–61]. Although there was little literature on bio-imprinting in sol–gel matrices, it was very effective to combine the mass-sensitive transducer with the surface imprinting technology and it can be used as an in situ analysis system for studying the interaction between the surfaces and biopolymers or even whole cells [62]. Additionally, in macroporous polysiloxane (silica) scaffolds, the imprinted protein can be prepared by changing the amount of protein loaded into the sol, and the quantity of surface-accessible protein related to the number of potential binding sites was varied [55].

1.3.2.2 TiO_2 Matrices

Titanium alkoxides can be used as inorganic gel matrices as well as silicon alkoxides. MIPs also can be prepared through classic sol–gel polymerization. The inorganic-framework molecularly imprinted TiO_2/WO_3 nanocomposites with molecular recognitive photocatalytic activity were prepared by the sol–gel method with the template molecules of 2-nitrophenol and 4-nitrophenol, which exhibited higher stability and selectivity. Tetrabutylorthotitanate was a titanium source that can be used as a precursor to functional monomer (Fig. 1.6) [63].

There are many articles about the study of MIPs modified TiO_2 nanocomposites each year [64]. Although the application of TiO_2 is limited due to its strong nonselective oxidation of holes and hydroxyl radicals generated by UV irradiation, MIPs with TiO_2 matrix have been proved to improve the relative adsorption capacity, selectivity and mass transfer rate of analytes [64]. In addition, MIPs with TiO_2 matrices have the advantages of high oxidation efficiency, nontoxicity, high photostability, chemical inertness, low cost, high binding affinity, chemical stability,

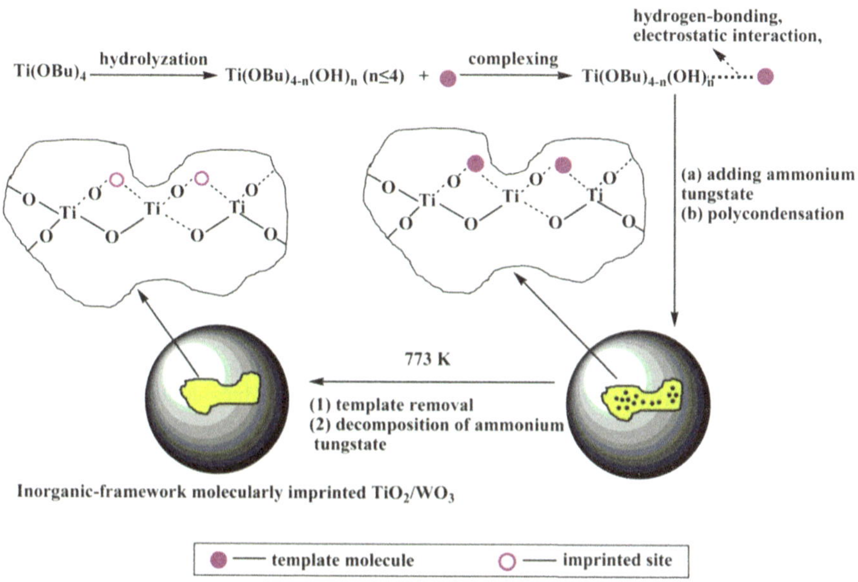

Fig. 1.6 Route for preparation of inorganic–framework molecularly imprinted TiO_2/WO_3 nanocomposite. (Reproduced with permission of Ref. [63])

easy way of preparation at a large scale, structure–activity predictability, specific recognition and wide practicability, and have been widely applied in the field of sensor construction, separation process, pollutant removal and drug development [65]. However, it also has some disadvantages, such as low utilization of visible-light, rapid recommendation of photogenerated electron/hole pairs, limitation and heterogeneity of the binding sites, cross-selectivity, template leakage and limited application in biology [64].

The methods of integrating of TiO_2-based nanomaterials into MIPs are generally divided into surface imprinting, precipitation polymerization and in situ polymerization [64]. Among them, surface imprinting is the main method. Surface molecular imprinting technique (SMIT), including graft copolymerization [66], sacrificial carrier method [67], sol–gel method [68], sol-hydrothermal polymerization [69], etc., is the major method and used to develop a molecular recognition system, so that specific target molecules have suitable binding sites on the substrate surface. This provides highly uniform size and shape for MIPs with TiO_2 nanomaterials. The method of precipitation polymerization (heterogeneous solution polymerization) includes liquid deposition method [70] and seed precipitation polymerization [71]. The functional monomers, cross-linkers, and initiators used in polymerization are dissolved in dispersants to form a uniform mixed solution. Although the application of this method is less than that of SMIT, the MIPs prepared by this method do not need surfactants and stabilizers, so the surface is clean and nonselective adsorption of imprinted molecules is avoided [72]. In situ polymerization is used for synthesizing the MIPs solid phase for chromatographic separation in the column using mixing solvent, template molecule, functional monomer, cross-linker and initiator [73]. In comparison to the other methods, the advantage of this method is that there is only one step reaction without the steps of grinding, sieving and sedimentation. However, it has not been explored enough.

1.3.2.3 Imprinting in Zeolites

Zeolite is a kind of crystalline aluminosilicate with definite pores and cavities throughout [36]. Because these metal oxide materials are not amorphous, the pore-size distribution is very uniform which is fixed by the atomic arrangements of their unit cells. The presence of nonvarying pore diameters in the small molecular scale endows the molecular sieves with extraordinary molecular identification ability [74]. Zeolites exhibit many properties that allow them to be used in the imprinting procedure. Subsequently, in the presence of template molecules, diethyl *p*-phenylenediacrylate (EPA), organic analogues of zeolites, with tunable molecular adsorption characteristics was synthesized by solid-state photopolymerization [75]. The vapor selectivity and molecular porosity of the particulate coatings were investigated in situ on the piezoelectric substrate by measuring sorption isotherms. With the decrease of template molecular size, smaller alkane molecules were gradually excluded from the MIP coating [75].

The structure of Zeolite is composed of tetrahedral $Si(Al)O_4$ units which are covalently linked by bridging O atoms to form more than 150 different types of frameworks [76]. To incorporate transition metal ions and organic units in the pores and make them become a part of zeolite framework, the materials of zeolitic imidazolate frameworks (ZIF) were developed, in which all tetrahedral atoms were transition metals, and all bridging ones were imidazolate units [76]. Subsequently, MIP composited ionic liquid-base dzeolitic imidazolate framework-67 was synthesized and using ZIF-67, ethylene glycol dimethacrylate as the cross-linker, benzoyl peroxide as the initiator, and heptane as the porogen to remove aristolochic acid I from herbal plant. Compared with other methods, the separation efficiency and selectivity were significantly improved [77].

In addition to the analytical field, the material has applications in related fields, such as drug delivery. A novel fluorescent zeolitic imidazolate framework-8-nanoparticles with doxorubicin (FZIF-8/DOX) as the core and MIP as the shell was designed, and the tumor sensitive biodegradable FZIF-8/DOX-MIP nanoparticles (FZIF-8/DOX-MIPs) were synthesized [78]. The MIP based on CD59 cell membrane glycoprotein epitope can recognize MCF-7 cancer cells (CD59 positive) and enrich fzif-8/DOX-MIPS to tumor site. *N,N*-′-diacrylylcystamine and dimethylaminoethyl methacrylate were used as the cross-linker and the monomer, respectively. The framework of MIP was broken under the stimulation of tumor microenvironment. The internal FZIF-8/DOX was exposed to microacidic environment, and DOX was released by further degradation. More importantly, FZIF-8/DOX-MIPs can be used not only in tumor treatment, but also in tumor diagnosis.

1.3.3 Magnetic Matrix

Magnetic molecularly imprinted polymers (MMIPs) originated in 1998 [79] were designed to improve the properties and applications in various fields. MMIPs are composed of magnetic materials, including iron, nickel, cobalt, or their alloys and oxides, as well as non-magnetic polymer materials, which have magnetic adsorption characteristics, high adsorption capacity and special selective recognition ability [34]. Fe_3O_4 is the most commonly used magnetic material due to its easy fabrication, low toxicity, and the abundant surface hydroxyls for modification [80].

The synthesis of MMIPS is similar to that of traditional MIPs. In the synthesis process, iron oxide (Fe_3O_4) called magnetite is introduced into polymerization to produce MMIPs [81]. The magnetic MIPs are usually composed of a magnetic core and a MIP layer at the shell, which have magnetic sensitivity and good selectivity to target compounds (Fig. 1.7) [82, 83]. However, in the previous studies, the magnetite surface needs to be modified before polymerization. For this purpose, a layer of SiO_2 was coated on magnetite by the reaction of TEOS. The obtained silanol groups were modified by γ-MAPS [84]. Another method is to modify the Fe_3O_4 with oleic acid or ethylene glycol to make the polymer surface amphoteric [81]. Then, MMIPs can be obtained by conventional synthesis method of MIPs. In addition, polychloromethylstyrene and chitosan can also be used for surface modification of the magnetite before polymerization [85].

Early MMIPS were designed as spheres. Then, MMIPs of nanotubes, nanocapsules, nanowires, and nanoparticles with high surface-to-volume ratios were developed to increase the binding capacity and kinetics [86]. On these bases, some novel configurations of MMIPs, including MMIPs with glassy carbon electrode [87], MMIPs with multi-walled carbon nanotube, MMIPs with nanosheet [88], hollow porous MMIPs [89], core-shell MMIPs [90], thermal-responsive MMIPs [91] were designed and developed to overcome the problems of traditional MIPs, such as template leakage, slow mass transfer rate, small binding capacity and poor site accessibility, and different sample requirements.

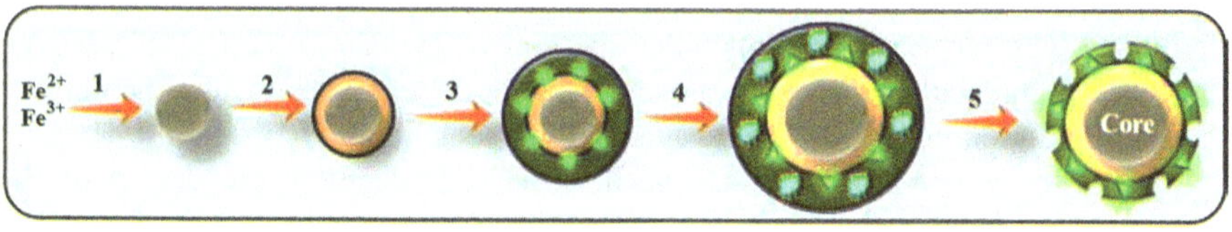

Fig. 1.7 Schematic representation of the preparation of magnetic molecular imprinted polymer. (1) Synthesis of magnetic core particle; (2) Silanization by tetraethoxy silane (TEOS); (3) Surface modification with 3-N-morpholinopropanesulfonic acid (MOPS); (4) Encapsulation and polymerization of MIP layer; (5) Template removal. (Reproduced with permission of Ref. [83])

1.4 Key Issues for a Rational Design

Typical MIP synthesis protocols include templates, functional monomers, cross-linkers, initiators and solvent (porogen). Due to the influence of the type and amount of monomer, crosslinking agent, initiator and solvent, reaction temperature and time, many attempts have been made to prepare MIPS with excellent properties. It is well known that the "three elements" of molecular imprinting include template molecules, functional monomers and cross-linkers, which are particularly worth studying.

An optimal polymerization experiment is the most important subject for the development of MIPs. But due to the influence of various factors, such as template, the type and amount of functional monomer, cross-linker, initiator, the temperature and time of polymerization reaction, the synthesis process of MIPs is full of challenges. In order to maintain the structure of the cavity after the template is removed, the polymers should be rather rigid, while a high flexibility of the polymers should be present to facilitate a fast equilibrium between the release and reuptake of the template in the cavity, which are contradictory to each other. It also requires as many cavities as possible, high thermal and mechanical stability [36]. Most studies are based on macroporous polymers with a high inner surface area (100–600 m^2/g) that show good accessibility. But for synthetic purposes, several more important considerations have to be emphasized to bring the information together according to the emerged rules.

1.4.1 Template

In all the molecular imprinting process, template is the most important, which guides the organization of functional groups based upon functional monomer [92]. However, not all templates can be directly used for templating. If the template is unstable in radical reactions, the alternative method has to be found [92]. The designed template needs to interact with the functional groups in the polymer. Due to the efforts of many researchers, in the initial covalent binding or subsequent non-covalent binding, many different types of templates and a large number of different binding site interactions are developed [36]. In addition, the concentration of template for non-covalent imprinting is also very important and should be optimized with respect to the functional monomer [93]. But, it is not necessary in the covalent molecular imprinting process because the template dictates the number of functional monomers that can be attached in a stoichiometric manner [94].

Moreover, two important issues need to be considered in the design of the template. One is the selection, which depends on the orientation of the functional groups in the cavity and the shape of the cavity [95, 96]. If the template has two binding sites, one two-point binding couple with several single-point binding can occur [25]. But only two-point binding is efficient. The portion of two-point binding can be increased through raising the temperature [97].

Another important factor is the rebinding of template molecules in the MIP cavity. In the case of covalent binding, the template determines the number of functional monomers that can be stoichiometric attached [94]. After the removal of original templates, binding sites are only located inside the cavity. This usually leads to a swelling of the cavities due to a salvation of the binding sites, which ensures a high reuptake percentage (90–95%) after the first removal [25]. At the same time, in the equilibrium process of template and polymer, it is helpful for rapid mass transfer. After reabsorbing the template, the cavity shrinks to its original volume (induced fit) [25].

In the usual non-covalent interaction, however, association constants are very low. Therefore, when the molar ratio is 1:1, only a small part of the template is combined. Only multiple hydrogen bonds show higher association constants [36]. Through the evaluation of several polymers with different formulations, the best ratio of template and functional monomer can be obtained [94]. Non-covalent imprinting, such as acrylic acid, usually requires at least four times more moles of binding sites to fully saturate the functionalities at the template molecule to ensure high selectivity. Therefore, the binding sites are not only distributed inside the cavity, but also in the whole polymer [36].

Sometimes, dual-template imprinting method would be introduced in the study. For example, traditional single template MIPs which are often used for either targeting or drug delivery are limited by the type of imprinted site. To cope with this problem, a dual-template MIP nanoparticle was prepared and applied to drug delivery and fluorescence imaging for BxPC-3 tumors based on their targeting ability. In this study, the 71–80 peptide of human fibroblast growth factor-inducible 14 modified by glucose (Glu-FH) and bleomycin (BLM) was used as templates simultaneously to enable FH-MIPNPs to load BLM and bind to the BxPC-3 cells. These imprinted sites enable the FH-MIPNPs to target fluorescence imaging and targeted therapy for specific tumors (Fig. 1.8) [98].

1.4.2 Functional Monomer

Due to the number of functional monomers used in molecular imprinting is limited, the selectivity and the further applications of MIPs are restricted to some extent. Therefore, the rapid development of more functional units is important for the development of MIT.

Functional monomers are responsible for the binding interactions of the imprinted binding sites. To obtain the maximized complex formation with the imprinting effect, it is obviously important to match the functionality of template and functional monomer in a complementary way (Fig. 1.9a) [92]. There must be at least one binding interaction, and stronger binding interaction(s) are better. However, the procedure of the functional monomer selection is very complex and difficult due to the multi-component systems, the time and resources required for polymer combination screening [99]. Combinatorial synthesis and computer simulation are

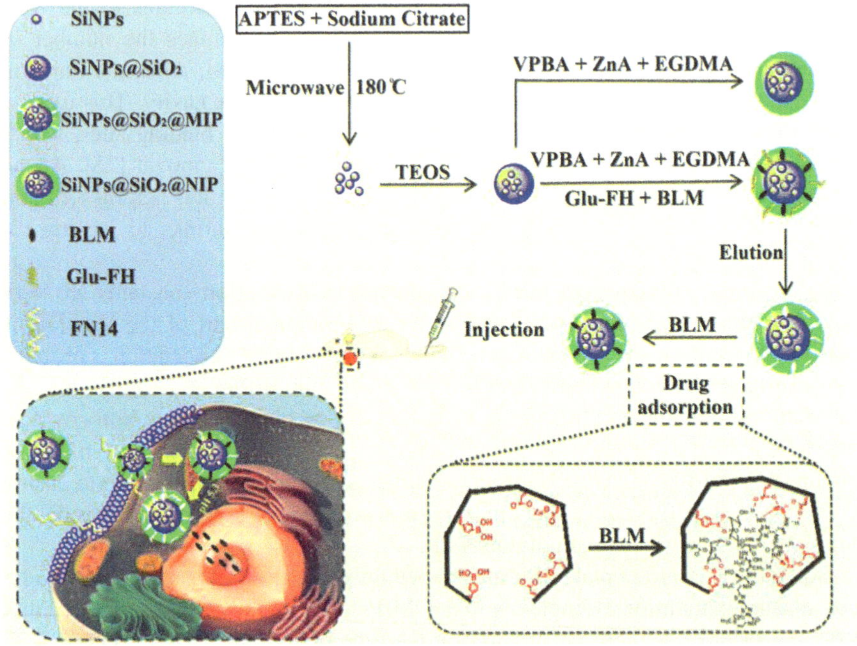

Fig. 1.8 Schematic diagram of the preparation and application of FH-MIPNPs. (Reproduced with permission of Ref. [98])

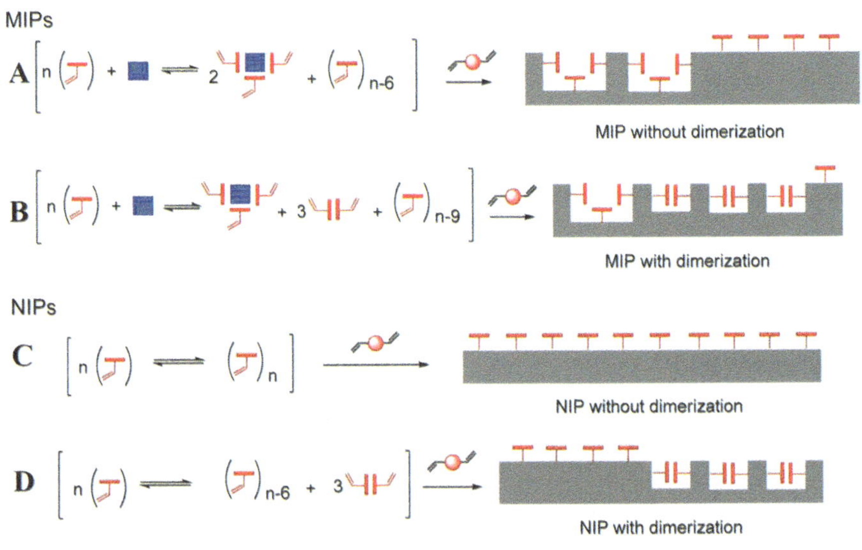

Fig. 1.9 Illustration comparing imprinted (**a**, **b**) and non-imprinted polymers (**c**, **d**) formed from functional monomers that have or lack the ability to dimerize. (Reproduced with permission of Ref. [108])

the most efficient methods to selected appropriate functional monomers [100–102], which permit the creation of MIPs with affinities and selectivities [103].

Combinatorial synthesis can quickly select a large number of products in a small scale to optimize the composition of MIP [104]. In the combinatorial method, the interaction between each monomer and the template can be compared to choose the functional monomers according to the experimental results of a set of non-imprinted polymers [99]. Since no template is required in the polymerization mixture, the polymer can be used with a variety of other templates, so it is simpler and cheaper than the traditional methods. More generally, the selection of functional monomers can be achieved by experiments on a set of non-imprinted polymers.

In the practical application of computational simulation, Leapfrog algorithm is used to analyze the combination between template and monomer in virtual library. These virtual libraries contain the most commonly used functional monomers that can form non-covalent interaction with template molecules, so as to obtain the best candidate preparation with the highest combination score. Finally, the arrangement of functional monomers around the template in a solvation box is optimized using the method of simulated annealing process [105, 106]. In addition, the functional monomer and solvent can also be virtually screened by theoretical model. For example, the interactions between the template molecule paclitaxel (PTX) and seven functional monomers embracing methacrylic acid (MA), acrolein (AC), 4-vinylbenzoic acid (4VA), acrylonitrile (AN), 2-vinylpyridine (2VP), 2,6-bisacrylamide pyridine (BAP), and methyl methacrylate (MM) were systematically investigated adopting the density functional theory (DFT) method. The different binding sites between PTX and solvents including chloroform, acetone, ethanol, methanol, and acetonitrile were studied. The calculated solvent energies and the binding energies of template-monomer showed that the chloroform was the most favorite solvent for the molecular imprinting reaction of PTX. Furthermore, according to the results of binding energy, the monomer 4VA c and PTX combined in the form of intermolecular hydrogen bond, which showed the most stable structure in the monomers studied. The experimental results can provide valuable theoretical guidance for the efficient extraction of PTX by MIT [107].

Moreover, most of the monomer units do not compound with templates, but form a large number of background sites. Monomer dimerization may reduce the number of these background sites (Fig. 1.9b). The binding sites formed by dimerized monomer will be effectively deactivated because the recognition group is blocked by self association. Among functional monomers, methacrylic acid (MAA) is a kind of widely used functional monomer because of its hydrogen bond donor and acceptor characteristics. The dimerization of MAA is beneficial to the moderate enhancement of imprinting effect. Compared with the non-imprinted polymers (NIPs) polymerized in the absence of template in different polar solvents (Fig. 1.9c, d), the investigation of effects of MAA dimerization on the imprinting process indicated that the ability of MAA to form dimers greatly reduced the number of background sites, but did not reduce the efficiency of imprinting process. In addition, the high mole fraction of MAA leaded to the large pore size of the polymer, which further improved the binding ability of the polymer [108].

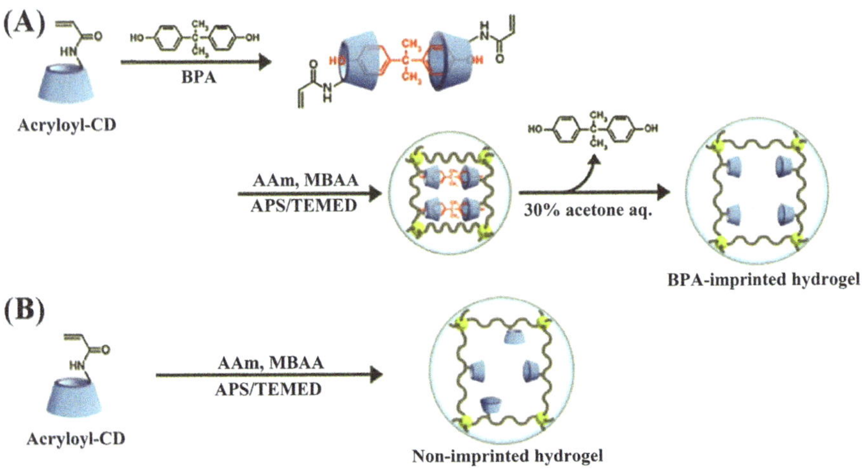

Fig. 1.10 Synthesis of BPA-imprinted (**a**) and non-imprinted hydrogel (**b**). (Reproduced with permission of Ref. [109])

Cyclic oligosaccharides such as β-cyclodextrins (β-CDs) have a hydrophilic appearance and a hydrophobic cavity. They can be used as functional monomers of MIT to form complexes with templates through hydrogen bonding, electrostatic interaction and host–guest interaction. In the presence of suitable crosslinking agents, the hydroxyl groups in the structure of β-CDs are important polymerization terminals for the formation of stable polymer matrix [35]. Figure 1.10 showed that a target molecule-responsive hydrogels with β-cyclodextrin (β-CD) were obtained by MIT using bisphenol A (BPA) as a template. The shrinkage of BPA-imprinted hydrogels was higher than that of non-imprinted hydrogels, because CD ligands arranged in suitable positions formed sandwich-like CD–BPA–CD complexes which played a cross-linking role [109].

Ionic liquids (ILs) also can be used as functional monomers. The structures of peptides with many interaction sites are very complex. To the high-efficiency imprinting peptides, ILs which provided multiple interactions with polypeptides, such as hydrogen bonding, π–π stacking, electrostatic interaction, hydrophobic interaction, coordination bonding, and van der Waals interaction, were introduced as functional monomers or surface modifiers. With the maturity of ILs as functional monomer or surface modifier, the interaction mechanism between MIPS and peptides based on ILs has been further explored. Ionic liquids usually contain organic cations and inorganic anions. Different types of ILs have different functions. For organic cations, imidazole ring and terminal functional groups can form a variety of interactions with dissolved biomolecules, such as π–π stacking, coordination bond, hydrogen bond, electrostatic interaction and van der Waals interaction [110]. The adsorption recognition ability can be improved by adjusting the functional groups of organic cations. Figure 1.11a showed the recognition performance of synthesized IL-functionalized microspheres with a series of different terminal functional groups.

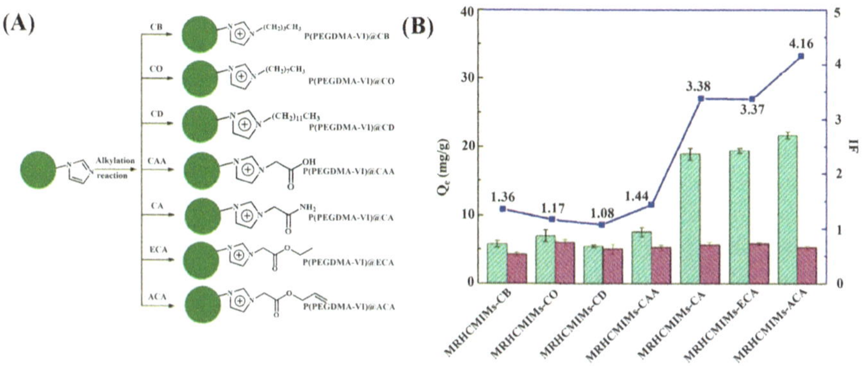

Fig. 1.11 Schematic illustration of different ILs as surface modifiers (**a**); results of the recognition performance of different IL-functionalized MIMs (**b**). (Reproduced with permission of Ref. [111])

By comparison, the MIPs prepared by allyl chloroacetate functionalized microspheres exhibited the best adsorption performance and imprinted factor (Fig. 1.11b) [111]. The properties of imprinted sites can be changed by the C=C bond of alkenyl ILs, which is very important for the excellent adsorption and recognition properties. For inorganic anions, ILs based on PF_6^- and $CF_3SO_3^-$ showed a higher imprinting effect than traditional Cl^- or BF_4^-, because PF_6^- was less nucleophilic, while $CF_3SO_3^-$ possessed a higher steric effect [112].

1.4.3 Cross-Linker

In the synthesis process, the cross-linkers play an important role in controlling the morphology, stabilizing the imprinted binding site and enhancing the mechanical stability of the polymer matrix [92]. A number of commercial cross-linkers can be used for the preparation of MIPs. Some of them can be complexed with the template as functional monomers simultaneously [92].

In the polymerization reaction, the reactivity ratios of cross-linking agent should match the reactivity rate of functional monomer to ensure the smooth combination of comonomers [92]. The selectivity mainly depended on the type and amount of cross-linker [93]. In the racemic resolution, cavities were not stable enough when the cross-linking in the polymer was less than a certain amount (approximately 10%), which leaded to insufficient selectivity. But the selectivity increased steadily with the increase of cross-linking. When the cross-linking degree was between 50% and 60%, the selectivity increased dramatically. As a classical cross-linking agent, ethylene dimethacrylate has become the most popular crosslinking agent in this field. Although cross-linking with divinylbenzene leads to the lower selectivity, the chemical stability is high because the bond is not hydrolyzable and has less interaction with functional groups [113].

1.4.4 Solvent

In the synthesis process, in addition to the introduction of template molecules, functional monomers and cross-linkers into one phase in the traditional method, the solvent also plays an important role in the formation of the porous structure of MIPs as the porogen [114]. The morphology of porosity and surface area and the total pore volume can be controlled by the nature and level of the porogen. The pore structures and specific surface areas could be improved by a thermodynamically excellent solvent and weakened by a thermodynamically poor solvent [92]. The volume of pore is positively correlated with the volume of porogen [92]. Additionally, the solvent can affect the complexation of functional monomers with the template in the whole polymerization process [92]. The shape and distance parameters of MIPs are determined by the solvation properties of solvents [94]. Chloroform is a widely used solvent because it can dissolve most of the templates and functional monomers, and not suppress the formation of hydrogen bonds. However, it should be noted that when the commercial chloroform was used as the solvent, the ethanol must be removed first, because it will prevent the formation of hydrogen bond between functional monomer and template, which is unfavorable for many molecular imprinting [30].

Recently, room temperature ionic liquids (RTILs), including [BMIM][BF_4], [BMIM][PF_6], [HMIM][PF_6], and [OMIM][PF_6], with unique characteristics have been used as solvents for the preparation of MIPs. The negligible vapor pressure of RTILs not only helps to reduce MIP bed shrinkage, but also acts as a pore template in polymerization. In addition, RTILs can also accelerate the synthesis process and improve the selectivity, adsorption and rebinding capacities [115].

1.4.5 Temperature

Temperature also affects the formation of MIPs. Compared with high temperature polymerization, the low temperature MIPs produced by photochemical polymerization have greater selectivity, which may be because low temperature helps to increase the number and the quality of binding sites [116, 117].

References

1. Gagliardi M, Mazzolai B (2015) Molecularly imprinted polymeric microand nano-particles for the targeted delivery of active molecules. Future Med Chem 7:123–138
2. Ye L (2013) Molecular imprinting: principles and applications of micro-and nanostructured polymers. CRC Press, FL
3. Alvarez-Lorenzo C, Concheiro A (2013) Handbook of molecularly imprinted polymers. A Smithers Group Company, Shropshire

4. BelBruno JJ (2019) Molecularly imprinted polymers. Chem Rev 119:94–119
5. Lehn JM (1995) Supramolecular chemistry, concepts and perspectives. Wiley, Weinheim
6. von Behring E, Kitasato S (1890) Ueber das Zustandekommen der Diphtherie-Immunitat und der Tetanus-Immunitat bei thieren. Deut Med Wochensch 16:1113–1114
7. Ehrlich P (1891) Experimentelle Untersuchungen über Immunität. II. Ueber Abrin. Deut Med Wochensch 17:1218–1219
8. Ehrlich P (1900) Croonian lecture—on immunity with special reference to cell life. Proc R Soc Lond 66:424–448
9. Polyakov MV (1931) Adsorption properties and structure of silica gel. Zh Fiz Khim 2:799–805
10. Pauling L (1940) A theory of the structure and process of formation of antibodies. J Am Chem Soc 62:2643–2657
11. Pauling L, Campbell DH (1942) The production of antibody in vitro. Science 95:440–441
12. Burnet FM (1957) A modification of Jerne's theory of antibody production using the concept of clonal selection. Aust J Sci 20:67–69
13. Dickey FH (1949) The preparation of specific adsorbents. Proc Natl Acad Sci U S A 35:227–229
14. Curti R, Colombo U (1952) Chromatography of stereoisomers with "tailor made" compounds. J Am Chem Soc 74:3961
15. Haldeman RG, Emmett PH (1955) Specific adsorption of alkyl orange dyes on silica gel. J Phys Chem 59:1039–1043
16. Beckett AH, Anderson P (1957) A method for the determination of the configuration of organic molecules using "stereo-selective adsorbents". Nature 179:1074–1705
17. Bartels H (1974) B. Prijs in advances in chromatography. CRC Press, FL
18. Wulff G, Sharhan A (1972) Über die anwendung von enzymanalog gebauten polymeren zur racemattrennung. Angew Chem 84:364
19. Komiyama M, Takeuchi T, Mukawa T, Asanuma H (2003) Molecular imprinting from fundamentals to applications. Wiley, Weinheim
20. Wulff G, Sarhan A, Zabrocki K (1973) Enzyme-analogue built polymers and their use for the resolution of racemates. Tetrahedron Lett 14:4329–4332
21. Wulff G, Zabrocki K, Hohn J (1978) Optically active polyvinyl compounds with chirality in the main chain. Angew Chem Int Ed 17:535–536
22. Wulff G, Vesper W, Grobe-Einsler R, Sarhan A (1977) Enzyme-analogue built polymers, 4. on the synthesis of polymers containing chiral cavities and their use for the resolution of racemates. Die Makromol Chem 178:2799–2816
23. Arshady R, Mosbach K (1981) Synthesis of substrate-selective polymers by host-guest polymerization. Die Makromol Chem 182:687–692
24. Vlatakis G, Andersson LI, Müller R, Mosbach K (1993) Drug assay using antibody mimics made by molecular imprinting. Nature 361:645–647
25. Wulff G (1995) Molecular imprinting in cross-linked materials with the aid of molecular templates—a way towards artificial antibodies. Angew Chem Int Ed 34:1821–1832
26. Alexander C, Davidson L, Hayes W (2003) Imprinted polymers: artificial molecular recognition materials with applications in synthesis and catalysis. Tetrahedron 59:2025–2057
27. Poma A, Turner APF, Piletsky SA (2010) Advances in the manufacture of MIP nanoparticles. Trends Biotechnol 28:629–637
28. Whitcombe MJ, Rodriguez ME, Villar P, Vulfson EN (1995) A new method for the introduction of recognition site functionality into polymers prepared by molecular imprinting: synthesis and characterization of polymeric receptors for cholesterol. J Am Chem Soc 117:7105–7111
29. Zaidi SA (2016) Molecular imprinted polymers as drug delivery vehicles. Drug Deliv 23:2262–2271

30. Tan T (2010) Molecular imprinting technology and application. Chemical Industry Press, Beijing
31. Mayes AG, Whitcombe MJ (2005) Synthetic strategies for the generation of molecularly imprinted organic polymers. Adv Drug Deliv Rev 57:1742–1778
32. Schirhagl R (2014) Bioapplications for molecularly imprinted polymers. Anal Chem 86:250–261
33. Sharma PS, Dabrowski M, D'Souza F, Kutner W (2013) Surface development of molecularly imprinted polymer films to enhance sensing signals. Trends Anal Chem 51:146–157
34. Ansari S, Karimi M (2017) Recent configurations and progressive uses of magnetic molecularly imprinted polymers for drug analysis. Talanta 167:470–485
35. Chen L, Wang X, Lu W, Wu X, Li J (2016) Molecular imprinting: perspectives and applications. Chem Soc Rev 45:2137–2211
36. Wulff G (2002) Enzyme-like catalysis by molecularly imprinted polymers. Chem Rev 102:1–27
37. Alexander C, Andersson HS, Andersson LI, Ansell RJ, Kirsch N, Nicholls IA, O'Mahony J, Whitcombe MJ (2006) Molecular imprinting science and technology: a survey of the literature for the years up to and including 2003. J Mol Recognit 19:106–180
38. Shea KJ, Sasaki DY (1991) An analysis of small-molecule binding to functionalized synthetic polymers by ^{13}C CP/MAS NMR and FT-IR spectroscopy. J Am Chem Soc 113:4109–4120
39. Takano E, Taguchi Y, Ooya T, Takeuchi T (2012) Dummy template-imprinted polymers for bisphenol a prepared using a Schiff base-type template molecule with post-imprinting oxidation. Anal Lett 45:1204–1213
40. Ramstrom O, Nicholls IA, Mosbach K (1994) Synthetic peptide receptor mimics: highly stereoselective recognition in non-covalent molecularly imprinted polymers. Tetrahedron: Asymmetry 5:649–656
41. Lofgreen JE, Ozin GA (2014) Controlling morphology and porosity to improve performance of molecularly imprinted sol–gel silica. Chem Soc Rev 43:911–933
42. Tian D (2017) Molecularly imprinted polymer functional materials. Science Press, Beijing
43. Yan H, Row KH (2006) Characteristic and synthetic approach of molecularly imprinted polymer. Int J Mol Sci 7:155–178
44. Katz A, Davis ME (2000) Molecular imprinting of bulk, microporous silica. Nature 403:286–289
45. Lee SW, Ichinose I, Kunitake T (1998) Molecular imprinting of azobenzene carboxylic acid on a TiO_2 ultrathin film by the surface sol-gel process. Langmuir 14:2857–2863
46. Hunnius M, Rufinska A, Maier WF (1999) Selective surface adsorption versus imprinting in amorphous microporous silicas. Microporous Mesoporous Mater 29:389–403
47. Pinel C, Loisil A, Gallezot P (1997) Preparation and utilization of molecularly imprinted silicas. Adv Mater 9:582–585
48. Gupta R, Kumar A (2008) Molecular imprinting in sol–gel matrix. Biotechnol Adv 26:533–547
49. Dai S, Shin YS, Barnes CE, Toth LM (1997) Enhancement of uranyl adsorption capacity and selectivity on silica sol–gel glasses via molecular imprinting. Chem Mater 9:2521–2525
50. Hench LL, West JK (1990) The sol–gel process. Chem Rev 90:33–72
51. Gupta R, Kumar A (2008) Bioactive materials for biomedical applications using sol–gel technology. Biomed Mater 3:034005
52. Tripathi VS, Kandimalla VB, Ju H (2006) Preparation of ormosils and its applications in the immobilizing biomolecules. Sens Actuators B 114:1071–1082
53. Arkhireeva A, Hay JN, Manzano M (2005) Preparation of silsesquioxane particles via a nonhydrolytic sol–gel route. Chem Mater 17:875–880
54. Han DM, Fang GZ, Yan YP (2005) Preparation and evaluation of a molecularly imprinted sol–gel material for on-line solid-phase extraction coupled with high performance liquid chromatography for the determination of trace pentachlorophenol in water samples. J Chromatogr A 1100:131–136

55. Lee K, Itharaju RR, Puleo DA (2007) Protein-imprinted polysiloxane scaffolds. Acta Biomater 3:515–522
56. Yang HH, Zhang SQ, Yang W, Chen XL, Zhuang ZX, Xu JG, Wang XR (2004) Molecularly imprinted sol-gel nanotubes membrane for biochemical separations. J Am Chem Soc 126:4054–4055
57. Kunitake T, Lee SW (2004) Molecular imprinting in ultrathin titania gel films via surface sol–gel process. Anal Chim Acta 504:1–6
58. Li F, Li J, Zhang SS (2008) Molecularly imprinted polymer grafted on polysaccharide microsphere surface by the sol–gel process for protein recognition. Talanta 74:1247–1255
59. Martin CR (1994) Nanomaterials: a membrane-based synthetic approach. Science 266:1961–1966
60. Jirage KB, Hulteen JC, Martin CR (1997) Nanotubule-based molecular-filtration membranes. Science 278:655–658
61. Miller SA, Young VY, Martin CR (2001) Electroosmotic flow in template-prepared carbon nanotube membranes. J Am Chem Soc 123:12335–12342
62. Dickert FL, Hayden O (2002) Bioimprinting of polymers and sol–gel phases. Selective detection of yeasts with imprinted polymers. Anal Chem 74:1302–1306
63. Luo X, Deng F, Min L, Luo S, Guo B, Zeng G, Au C (2013) Facile one-step synthesis of inorganic-framework molecularly imprinted TiO_2/WO_3 nanocomposite and its molecular recognitive photocatalytic degradation of target contaminant. Environ Sci Technol 47:7404–7412
64. Sun L, Guan J, Xu Q, Yang X, Wang J, Hu X (2018) Synthesis and applications of molecularly imprinted polymers modified TiO_2 nanomaterials: a review. Polymers (Basel) 10:1248
65. Daghrir R, Drogui P, Robert D (2013) Modified TiO_2 for environmental photocatalytic applications: a review. Ind Eng Chem Res 52:3581–3599
66. Roy E, Patra S, Madhuri R, Sharma PK (2016) A single solution for arsenite and arsenate removal from drinking water using cysteine@zns: TiO_2 nanoparticle modified molecularly imprinted biofouling-resistant filtration membrane. Chem Eng J 304:259–270
67. Xu WZ, Zhou W, Xu PP, Pan JM, Wu XY, Yan YS (2011) A molecularly imprinted polymer based on TiO_2 as a sacrificial support for selective recognition of dibenzothiophene. Chem Eng J 172:191–198
68. Liu Y, Liu R, Liu C, Luo S, Yang L, Sui F, Teng Y, Yang R, Cai Q (2010) Enhanced photocatalysis on TiO_2 nanotube arrays modified with molecularly imprinted TiO_2 thin film. J Hazard Mater 182:912–918
69. Deng F, Liu Y, Luo X, Wu S, Luo S, Au C, Qi R (2014) Sol-hydrothermal synthesis of inorganic-framework molecularly imprinted TiO_2/SiO_2 nanocomposite and its preferential photocatalytic degradation towards target contaminant. J Hazard Mater 278:108–115
70. Maki H, Okumura Y, Ikuta H, Mizuhata M (2014) Ionic equilibria for synthesis of TiO_2 thin films by the liquid-phase deposition. J Phys Chem C 118:11964–11974
71. Rauh A, Honold T, Karg M (2016) Seeded precipitation polymerization for the synthesis of gold-hydrogel core-shell particles: the role of surface functionalization and seed concentration. Colloid Polym Sci 294:37–47
72. Li GL, Moehwald H, Shchukin DG (2013) Precipitation polymerization for fabrication of complex core-shell hybrid particles and hollow structures. Chem Soc Rev 42:3628–3646
73. Moein MM, Javanbakht M, Akbari-Adergani B (2011) Molecularly imprinted polymer cartridges coupled on-line with high performance liquid chromatography for simple and rapid analysis of dextromethorphan in human plasma samples. J Chromatogr B 879:777–782
74. Davis ME, Katz A, Ahmad WR (1996) Rational catalyst design via imprinted nanostructured material. Chem Mater 8:1820–1839
75. Guney-Altay O, Pestov D, Tepper G (2007) Organic zeolites from a diolefinic monomer. J Am Chem Soc 129:13957–13962

76. Park KS, Ni Z, Cote AP, Cho JY, Huang R, Uribe-Romo FJ, Chae HK, O'Keeffe M, Yaghi OM (2006) Exceptional chemical and thermal stability of zeolitic imidazolate frameworks. Proc Natl Acad Sci U S A 103:10186–10191
77. Fang L, Tian M, Row KH, Yan X, Xiao W (2019) Isolation of aristolochic acid I from herbal plant using molecular imprinted polymer composited ionic liquid-based zeolitic imidazolate framework-67. J Sep Sci 42:3047–3053
78. Qin YT, Feng YS, Ma YJ, He XW, Li WY, Zhang YK (2020) Tumor-sensitive biodegradable nanoparticles of molecularly imprinted polymers-stabilized fluorescent zeolitic imidazolate framework-8 for targeted imaging and drug delivery. ACS Appl Mater Interfaces 12:24585–24598
79. Ansell RJ, Mosbach K (1998) Molecularly imprinted polymers, man-made mimics of antibodies and their application in analytical chemistry. Analyst 123:1611–1616
80. Xie LJ, Jiang RF, Zhu F, Liu H, Ouyang GF (2014) Application of functionalized magnetic nanoparticles in sample preparation. Anal Bioanal Chem 406:377–399
81. Dramou P, Zuo P, He H, Pham-Huy LA, Zou W, Xiao D, Pham-Huy C (2013) Development of novel amphiphilic magnetic molecularly imprinted polymers compatible with biological fluids for solid phase extraction and physicochemical behavior study. J Chromatogr A 1317:110–120
82. Chen L, Li B (2012) Application of magnetic molecularly imprinted polymers in analytical chemistry. Anal Methods 4:2613–2621
83. Ashley J, Shahbazi MA, Kant K, Chidambara VA, Wolff A, Bang DD, Sun Y (2017) Molecularly imprinted polymers for sample preparation and biosensing in food analysis: progress and perspectives. Biosens Bioelectron 91:606–615
84. Bouri M, Lerma-García MJ, Salghi R, Zougagh M, Ríos A (2012) Selective extraction and determination of catecholamines in urine samples by using a dopamine magnetic molecularly imprinted polymer and capillary electrophoresis. Talanta 99:897–903
85. Fan L, Zhang Y, Li X, Luo C, Lu F, Qiu H (2012) Removal of alizarin red from water environment using magnetic chitosan with alizarin red as imprinted molecules. Colloids Surf B 91:250–257
86. Ansari S, Karimi M (2017) Novel developments and trends of analytical methods for drug analysis in biological and environmental samples by molecularly imprinted polymers. TrAC Trends Anal Chem 89:147–162
87. Rao H, Lu Z, Ge H, Liu X, Chen B, Zou P, Wang X, He H, Zeng X, Wang Y (2017) Electrochemical creatinine sensor based on a glassy carbon electrode modified with a molecularly imprinted polymer and a Ni@polyaniline nanocomposite. Microchim Acta 184:261–269
88. Ning F, Qiu T, Wang Q, Peng H, Li Y, Wu X, Zhang Z, Chen L, Xiong H (2017) Dummy-surface molecularly imprinted polymers on magnetic graphene oxide for rapid and selective quantification of acrylamide in heat-processed (including fried) foods. Food Chem 221:1797–1804
89. Li H, Hu X, Zhang YP, Shi SY, Jiang XY, Chen XQ (2015) Highcapacity magnetic hollow porous molecularly imprinted polymers for specific extraction of protocatechuic acid. J Chromatogr A 1404:21–27
90. Sheykhaghaei G, Hossaini-Sadr M, Khanahmadzadeh S (2016) Synthesis and characterization of core-shell magnetic molecularly imprinted polymer nanoparticles for selective extraction of tizanidine in human plasma. Bull Mater Sci 39:647–653
91. Ma W, Li S, Chen L, Sun J, Yan Y (2017) Core-shell thermal-responsive and magnetic molecularly imprinted polymers based on mag-yeast for selective adsorption and controlled release of tetracycline. J Iran Chem Soc 14:209–219
92. Cormack PAG, Elorza AZ (2004) Molecularly imprinted polymers: synthesis and characterization. J Chromatogr B 804:173–182
93. Kim H, Spivak DA (2003) New insight into modeling non-covalently imprinted polymers. J Am Chem Soc 125:11269–11275

94. Spivak DA (2005) Optimization, evaluation, and characterization of molecularly imprinted polymers. Adv Drug Deli Rev 57:1779–1794
95. Shea KJ, Sasaki DY (1989) On the control of microenvironment shape of functionalized network polymers prepared by template polymerization. J Am Chem Soc 111:3442–3444
96. O'Shannessy DJ, Andersson LI, Mosbach K (1989) Molecular recognition in synthetic polymers. Enantiomeric resolution of amide deriva-tives of amino acids on molecularly imprinted polymers. J Mol Recogn 2:1–5
97. Wulff G, Kirstein G (1990) Measuring the optical activity of chiral imprints in insoluble highly cross-linked polymers. Angew Chem Int Ed 29:684–686
98. Jia C, Zhang M, Zhang Y, Ma ZB, Xiao NN, He XW, Li WY, Zhang YK (2019) Preparation of dual-template epitope imprinted polymers for targeted fluorescence imaging and targeted drug delivery to pancreatic cancer BxPC-3 cells. ACS Appl Mater Interfaces 11:32431–32440
99. Karim K, Breton F, Rouillon R, Piletska EV, Guerreiro A, Chianella I, Piletsky SA (2005) How to find effective functional monomers for effective molecularly imprinted polymers. Adv Drug Deliv Rev 57:1795–1808
100. Lanza F, Sellergren B (1999) Method for synthesis and screening of large groups of molecularly imprinted polymers. Anal Chem 71:2092–2096
101. Herdes C, Sarkisov L (2009) Computer simulation of volatile organic compound adsorption in atomistic models of molecularly imprinted polymers. Langmuir 25:5352–5359
102. Nicholls IA, Andersson HS, Charlton C, Henschel H, Karlsson BCG, Karlsson JG, O'Mahony J, Rosengren AM, Rosengren KJ, Wikman S (2009) Theoretical and computational strategies for rational molecularly imprinted polymer design. Biosens Bioelectron 25:543–552
103. Whitcombe MJ, Chianella I, Larcombe L, Piletsky SA, Noble J, Porter R, Horgan A (2011) The rational development of molecularly imprinted polymer-based sensors for protein detection. Chem Soc Rev 40:1547–1571
104. Takeuchi T, Fukuma D, Matsui J (1991) Combinatorial molecular imprinting: an approach to synthetic polymer receptors. Anal Chem 71:285–290
105. Piletsky SA, Karim K, Piletska EV, Day CJ, Freebairn KW, Legge C, Turner APF (2001) Recognition of ephedrine enantiomers by molecularly imprinted polymers designed using a computational approach. Analyst 126:1826–1830
106. Subrahmanyam S, Piletsky SA, Piletska EV, Chen BN, Karim K, Turner APF (2001) "Bite-and-switch" approach using computationally designed molecularly imprinted polymers for sensing of creatine. Biosens Bioelectron 16:631–637
107. Wang L, Yang F, Zhao X, Li Y (2020) Screening of functional monomers and solvents for the molecular imprinting of paclitaxel separation: a theoretical study. J Mol Model 26:26
108. Zhang YG, Song D, Lanni LM, Shimizu KD (2010) Importance of functional monomer dimerization in the molecular imprinting process. Macromolecules 43:6284–6294
109. Kawamura A, Kiguchi T, Nishihata T, Uragami T, Miyata T (2014) Target molecule-responsive hydrogels designed via molecular imprinting using bisphenol A as a template. Chem Commun 50:11101–11103
110. Wan D, Yan C, Liu Y, Zhu K, Zhang Q (2019) A novel mesoporous nanocarrier: integrating hollow magnetic fibrous silica with PAMAM into a single nanocomposite for enzyme immobilization. Microporous Mesoporous Mater 280:46–56
111. Du C, Hu X, Guan P, Gao X, Song R, Li J, Qian L, Zhang N, Guo L (2016) Preparation of surface-imprinted microspheres effectively controlled by orientated template immobilization using highly cross-linked raspberry-like microspheres for the selective recognition of an immunostimulating peptide. J Mater Chem B 4:1510–1519
112. Ding HY, Chen RF, Liu MM, Huang R, Du YM, Huang C, Yu XY, Feng XH, Liu F (2016) Preparation and characterization of biocompatible molecularly imprinted poly (ionic liquid) films on the surface of multi-walled carbon nanotubes. Rsc Adv 6:43526–43538
113. Wulff G, Vietmeier J, Poll HG (1987) Influence of the nature of the crosslinking agent on the performance of Imprinted polymers in racemic resolution. Makromol Chem 188:731–740

114. Lloyd L (1991) Rigid macroporous copolymers as stationary phases in high-performance liquid-chromatography. J Chromatogr 544:201–217
115. Xu Z, Fang G, Wang S (2010) Molecularly imprinted solid phase extraction coupled to high-performance liquid chromatography for determination of trace dichlorvos residues in vegetables. Food Chem 119:845–850
116. O'Shannessy DJ, Ekberg B, Mosbach K (1989) Molecular imprinting of amino acid derivatives at low temperature (0 °C) using photolytic homolysis of azobisnitriles. Anal Biochem 177:144–149
117. Lanza F, Sellergren B (2004) Molecularly imprinted polymers via high-throughput and combinatorial techniques. Macromol Rapid Commun 25:59–68

Chapter 2
Synthetic Strategies for the Generation of Molecularly Imprinted Polymers

Xi Wu

2.1 Introduce

Three general approaches, covalent, semi-covalent and non-covalent, are frequently applied in the preparation of MIPs. For the covalent approach, the template is bound to a polymerizable monomer by a labile covalent bond. The template and monomer are copolymerized with a cross-linker in a selected porogenic solvent. A cross-linked polymer is formed around the template which is fixed in the MIPs via labile chemical bond. Then, the bond is broken to remove the template, leaving binding cavities which is uniform in placement of a complementary functional group. However, there are only a limited number of examples that utilize covalent approach in the literature. That is because this approach demands multiple heteroatom functionality to be available in the template.

Because of flexible application and less restrictions, non-covalent imprinting is the most popular imprinting strategy. In the non-covalent approach, the usual preparation process of MIPs could be summarized as polymerization of functional monomer and cross-linker around a template in porogenic solvent. The interactions between the template and MIPs usually are a combination of non-covalent interactions such as H-bonding, electrostatic or π–π interactions. Then, the template is eluted from the MIPs and recognition cavities complementary to the template in chemical functionality, shape and size are formed in the MIPs. These recognition cavities can specifically rebind the template or its analogues from a complex mixture. The main drawback of non-covalent approach is the non-selective binding sites obtained arising from the multitude of complexes with different template-monomer stoichiometry. The pre-polymerization step of non-covalent approach is far from a perfect defined process and it leads to an excess of functional monomer relative to the template exists in the polymerization system. These "free" functional

X. Wu (✉)
Department of Chemistry, Changzhi University, Changzhi, China

Z. Liu et al. (eds.), *Molecularly Imprinted Polymers as Advanced Drug Delivery Systems*, https://doi.org/10.1007/978-981-16-0227-6_2

monomers are randomly distributed in the polymers and it lead to the non-selective binding sites were formatted.

The third approach is called semi-covalent method and it combines the advantage of both covalent and non-covalent approaches. In the copolymerization step of this strategy, the template interacts with the functional monomer via reversible covalent bond, just as in the covalent approach. Then, the template is removed, leaving an imprint bearing functional groups. In the rebinding step, only non-covalent interactions are exploited, exactly as in the non-covalent approach. The semi-covalent approach is a practical method to obtain MIPs with higher specific recognition ability since it is a feasible way for controlled tailoring of homogeneous binding sites during polymerization.

2.2 Covalent Imprinting

Covalent imprinting approach is a typical method to produce MIPs and it could produce pretty homogeneous recognition sites which exist only in the imprinted cavities in polymer matrix. It demands the template is chemically bound to a monomer by means of a labile covalent interaction. Some reversible condensation reactions such as ketals/acetals, Schiff's base and boronate esters are often utilized to achieve this purpose. However, covalent imprinting is not a common method for the MIPs preparation since the type of reversible condensation reactions is limited.

2.2.1 Covalent Imprinting with Boronate Esters

In covalent imprinting, boronic acids could be used as monomer because the related groups are pretty fit for covalent binding. It can combine with diol-containing molecules to form fairly stable boronic esters. During the reaction, ester formation takes place at relatively low reaction rates and no back reaction occurs in the absence of water. Templates can be removed easily in alcohol or water because boronic esters can be hydrolyzed fastly and completely in these conditions. By contrast, the rebinding of template afterwards is quite slow. Obviously, the rate of equilibration is not satisfying. Luckily, in alkaline solution or in the presence of certain nitrogen bases, esterification reaction of boronic acid with diol occurred, which equilibrate extremely quickly [1, 2]. If the interaction with nitrogen bases takes place intramolecularly, the rate of equilibration would be further accelerated [3].

The boronic acids monomer is very suitable for the template molecule contain diol groups. The template can be bound to the monomer by the boronic ester bond and MIPs with high selectivity could be obtained if all boronic acid binding sites are sufficiently utilized. Wulff and coworker [4, 5] prepared the MIPs using various sugar racemates as templates and boronic acids as functional monomer. The

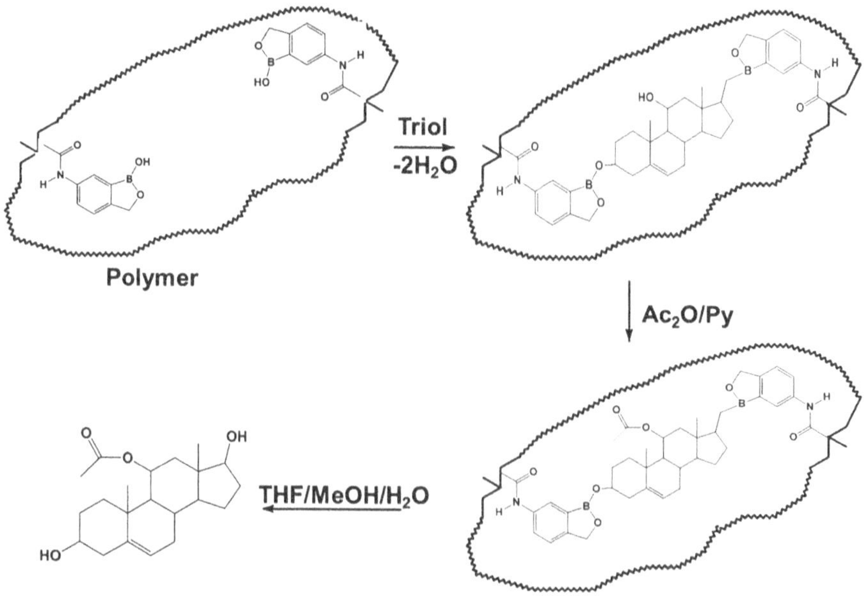

Fig. 2.1 Modification of androst-5-ene-3β,11β,17β-triol on polymer. (Reproduced with permission of Ref. [6])

obtained MIPs were able to resolve racemates of the templates and show a good performance. The boronic acid functional monomer can also be used for binding some other compounds, such as monoalcohols and steroid alcohols (Fig. 2.1) [2, 6].

2.2.2 Covalent Imprinting with Schiff's Bases

Many studies have shown Schiff's bases are also suitable for covalent imprinting [7–9]. The equilibrium rate of the Schiff's bases reaction can satisfy the needs of imprinting with the aid of catalysts or suitable intramolecular neighboring groups. In principle, the active site for covalent imprinting based on Schiff's bases is an amine or an aldehyde. For instance, the aldehyde containing binding sites can be used for the imprinting with amino acid derivatives [7, 8]. The MIPs obtained by covalent imprinting with Schiff's bases have superior selectivity for rebinding the template. That is because two amino groups could be introduced in polymer and two binding groups could provide substrate selectivity. In brief, the selectivity of MIPs obtained by this method is related to the arrangement of the functional groups within the cavity [9].

2.2.3 Covalent Imprinting with Ketals and Acetals

Ketals (Fig. 2.2) for covalent imprinting was also investigated [10–12]. In these researches, diketones with different distances between the keto groups react with polymerizable diols to prepare functional monomers. The MIPs obtained by this strategy showed excellent recognition selectivity with diketones as the template. With these diketones based monomers, the mechanism of the imprinting procedure was studied. The arrangement of functional groups is one major factor that affects performance of the polymer. Furthermore, the size and shape of the template molecules are also important factors for the separability. Similarly, acetals have been successfully employed for the covalent imprinting. A suitable cyclic half-acetal can form full acetals with monoalcohols and this reaction can be used for reversible binding of alcohols.

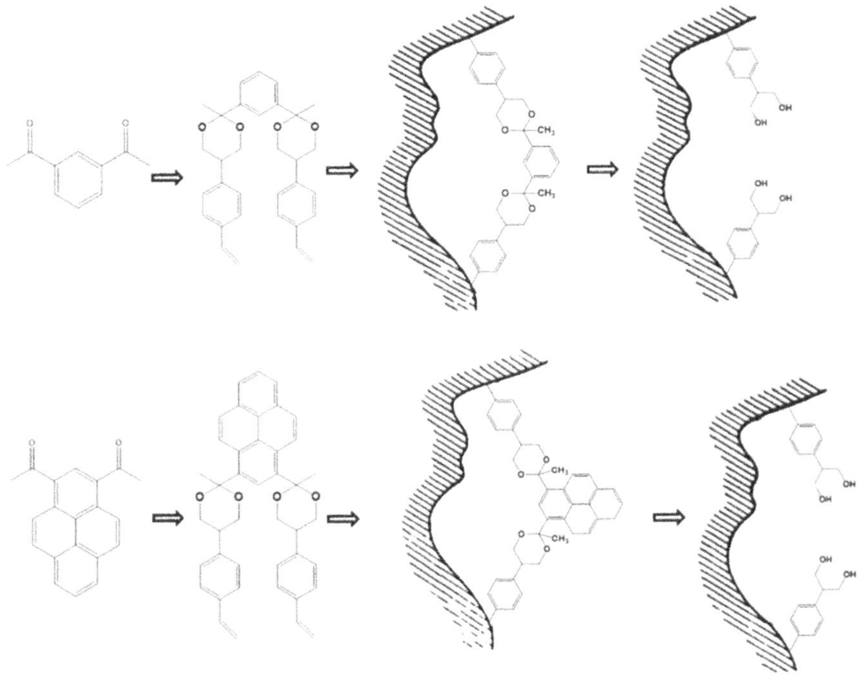

Fig. 2.2 Ketals for covalent imprinting. (Reproduced with permission of Ref. [11])

2.3 Semi-Covalent Imprinting

Both of covalent and non-covalent imprinting methods have its own advantage and disadvantage. Because the template reacts with a certain amount of functional monomer by covalent bond in the polymerization system, the covalent imprinting method remarkably reduced the nonspecific binding sites. It makes that all functional groups present an ideal situation for the imprinting process. However, the complex rebinding process makes covalent imprinting approach not practical for the most applications. Dehydration and hydrolysis reaction for template removal and rebinding template slow the interchange of template with the sites. Furthermore, covalent imprinting is fairly restrictive as it is a challenge to find an appropriate covalent template-monomer complex. Non-covalent imprinting is by far the most important MIPs preparation method. It has many advantages such as simplicity of the synthesis process and broad applicability to a wide range of template structures. The obtained MIPs show excellent properties, such as good mechanical property, high chemical stability and low cost. However, the non-covalent imprinting also has some drawbacks. Since the equilibrium nature of template-monomer interactions, the excess monomer is necessary to displace the equilibrium to form the template-monomer complex. Moreover, the pre-polymerization complexes are formed with an uncontrollable template: monomer stoichiometry. All of these factors can lead to nonspecific binding.

To overcome the problems with covalent and non-covalent approach, a hybrid approach, semi-covalent approach, is proposed. It combines the advantages of covalent and non-covalent method. During the semi-covalent polymerization, the template is bound to a monomer through a reversible covalent bond at first and it can be called template-monomer. After splitting of the template, this step usually achieved by hydrolysis, the generated groups become available to rebinding through non-covalent interactions [13, 14]. Because the imprinting process is covalent, the randomly distributed functional groups from free monomers are limited for the semi-covalent approach and all the binding sites are pretty uniform. Furthermore, template rebinding process is hardly affected by kinetic restrictions.

In the semi-covalent imprinting, template and monomer can be reacted by an ester or amide linkage. The research of Cheong et al. [15] can be considered as a suitable example for this method. At first, the template-monomer was synthesized via esterification of testosterone with methacryloyl chloride in the presence of triethylamine. Then, taken testosterone methacrylate as the template-monomer, EDMA as cross-linker, and chloroform as porogen, the MIPs was synthesized. The obtained MIPs was hydrolyzed with NaOH methanol solution followed by acidification. At last, methacrylic acid residues can be obtained in the polymer and was used for binding testosterone via hydrogen bonding.

For semi-covalent MIPs, removal of the template is achieved by template-monomer hydrolysis. However, template-monomer hydrolysis is usually not easy. Steric crowding is major obstacle in the rebinding process because the steric

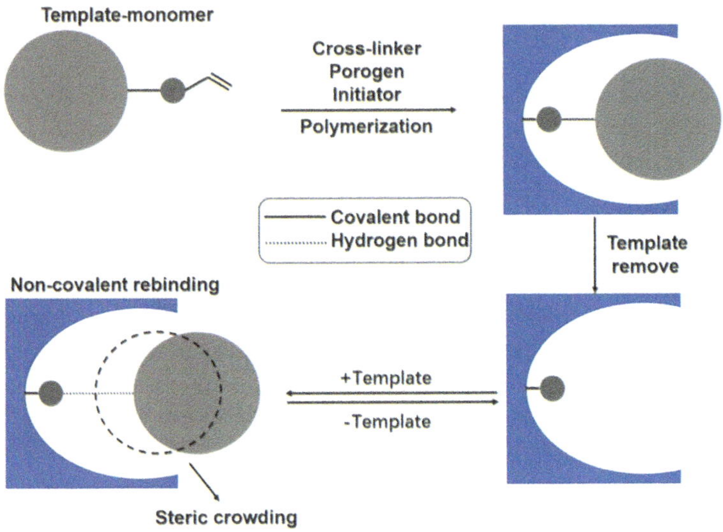

Fig. 2.3 Outline of the semi-covalent imprinting method

requirements of template to monomer in hydrogen bonding contact are different from the corresponding template-monomer (Fig. 2.3).

Hence, some researches about semi-covalent imprinting introduced a linker group which is lost on template removal between the template and the functional monomer to this difficulty [16]. This linker group is known as "spacer" because it works as a spacer between the template and polymer-bound functionality to avoid steric crowding in the process of rebinding.

Chen and coworkers [17] have also adopted semi-covalent imprinting with carbonyl group as sacrificial spacer to synthesis MIP for phenols. In this research, 4-chlorophenyl (4-vinyl)phenyl carbonate was taken as template-monomer, EDMA, 2,2-azobisisobutyronitrile (AIBN), and chloroform was applied as cross-linker, initiator, and porogen, respectively (Fig. 2.4). For removing the template, the MIPs were hydrolyzed using the strategy proposed by Whitcombe and coworkers [14]. The MIPs show superior selectivity for phenols and was successfully applied as HPLC stationary phase in the determination of phenols.

Several other groups can also be used as sacrificial spacers to prepare MIPs. In the research of Chang and coworkers [18], a molecularly imprinted spherical silica particles with controlled sizes was prepared. This material employed carbonyl spacer in template-monomer linked through carbamate linkage. In the presence of dibutyltin dilaurate, the template-monomer complex was synthesized by the reaction of estrone with 3-(triethoxysilyl)propyl isocyanate. Since the bond between template and silica monomer is thermally reversible, the template can be easily removed by thermal reaction.

Furthermore, the template and monomer could also be linked through carbonate. The cholesterol-MIPs were suggested by Lee and coworker [19]. The cholesteryl

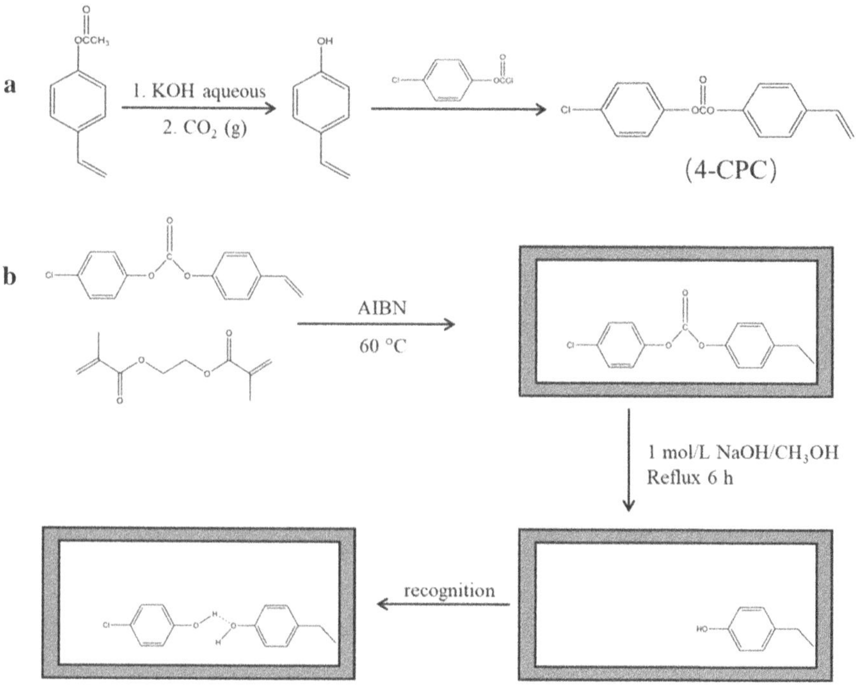

Fig. 2.4 Schematic of the sacrificial semi-covalent method applied in ref. 17. (**a**) The synthesis of template. (**b**) The preparation process of 4-chlorophenol MIPs by semi-covalent method. (Reproduced with permission of Ref. [17])

(4-vinyl) phenyl carbonate which was obtained by the approach of Whitcombe et al. [14] was applied as template-monomer. The MIPs was packed into the HPLC column as stationary phase and the column was used to isolate cholesterol from other steroids. The results also proved that the polymer had a good adsorption capacity for cholesterol.

A variant of the sacrificial spacer approach was employed to imprint pyridine and quinoline via the silyl esters and dimethyl silyl group of silyl ether [20]. In this study, silyl ether derivatized templates were designed for binding nitrogen heterocycles (Fig. 2.5a). The dimethyl silyl group of silyl ether and silyl esters could be acidic hydrolysis or nucleophilic displacement with fluoride ion under mild conditions (Fig. 2.5b). It is means that the templates can be removed easily. In addition, silyl ether chemistry condensed aromatic templates and improved the solubility of templates in nonpolar porogens in MIPs synthesis. The proposed method was successful in creating affinity in MIPs for the nitrogen heterocycles.

There is a method based on urea linkages to introduce amine groups into the MIPs [9]. In this kind of semi-covalent imprinting approach, urea linkage is formed by linking two nitrogen atoms with carbonyl group. The C=O group in urea linkages was applied as sacrificial spacer. In the research of Lubke et al. [21], the bis-N-(4-vinylphenyl)urea derivative of 2,8-dichloro-3,7-diaminodibenzodioxin, used the

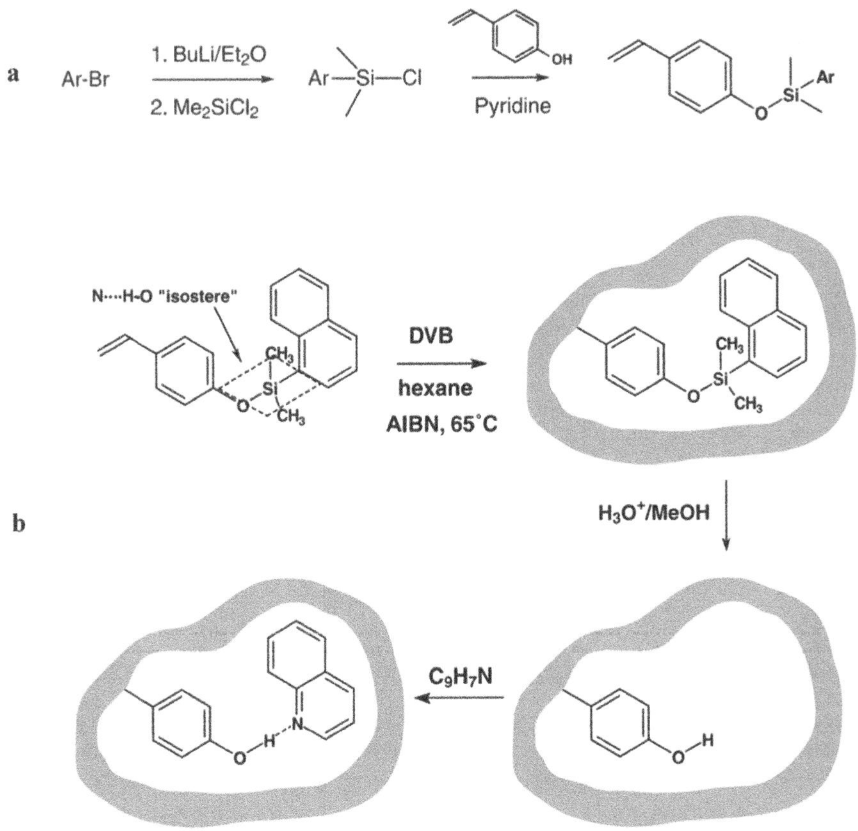

Fig. 2.5 (**a**) Preparation scheme for the synthesis of silyl ether templates. (**b**) Schematic of the imprinting process using a silyl ether as an N···H–O isostere and single atom sacrificial spacer to create recognition sites for nitrogen heterocycles. (Reproduced with permission of Ref. [20])

C=O group spacer to introduce aromatic amines into the MIPs after removal of the template (Fig. 2.6). As a result, MIPs prepared with the diurea template-monomer has very good recognition performance for 2,3,7,8-tetrachlorodibenzodioxin.

Sacrificial spacer gives the semi-covalent imprinting a unique advantage to avoid steric crowding. However, due to the complicated preparation procedure of template-monomer before MIPs polymerization, only a few compounds, such as cholesterol [14, 19], estrone [18], propofol [22, 23], menthol [24, 25], DDT [26], and 2,3,7,8-tetrachlorodibenzodioxin (TCDD) [21], could be imprinted by the semi-covalent method.

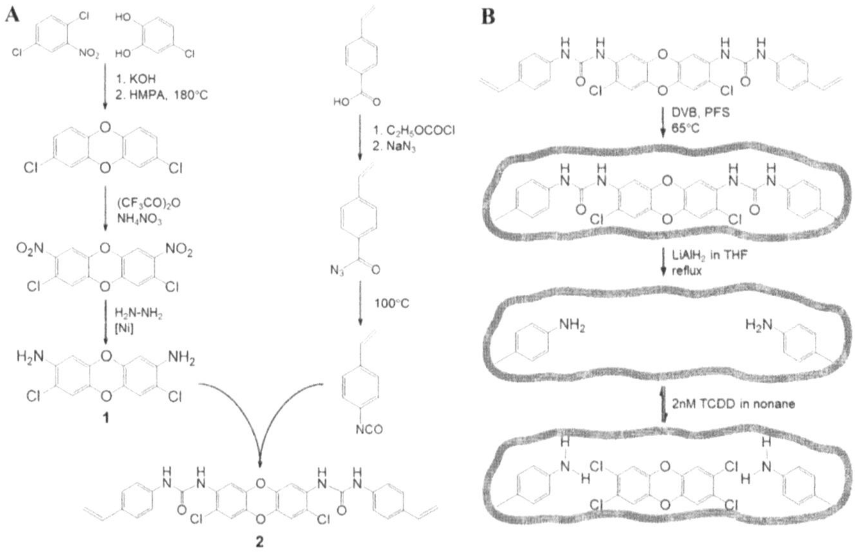

Fig. 2.6 (**a**) Preparation of the diurea template 2 and (**b**) Schematic of the synthesis the template 2 MIPs showing the positioning of aromatic amine groups in the recognition site through the urea functionality (incorporating a carbonyl spacer). (Reproduced with permission of Ref. [21])

2.4 Non-covalent Imprinting

Until now, non-covalent imprinting is the most widely used way for the preparation of MIPs. It is a kind of self-assembly approach and a simplest method to introduce functional groups into the cavities of polymer. The weak non-covalent intermolecular interactions of template-functional monomers have been employed for the template recognition. These are many types of non-covalent interactions and one of the most important is hydrogen bonding interaction due to the specific geometric directionality. The major weakness of non-covalent systems is the heterogeneous binding sites which are obtained from the non-well-defined pre-polymerization step. The formation of template-monomer complexes with different ratios leads to different binding sites, and excess monomers are used in order to form the complexes of template-monomer also might lead to non-selective binding sites [27].

2.4.1 The Nature of the Pre-polymerization Complex

In non-covalent imprinting, it is normally assumed that a pre-polymerization templates-functional monomer complex is formed before the polymerization reaction. Since templates and functional monomers might have multiple binding sites, the interactions between them could be pretty diversity. It is means that the pre-polymerization complex system is pretty complexity. The interactions between templates and functional monomers could be affected by some factors, such as the property of solvent (the polarity or hydrogen bonding strength) and the reaction temperature. Generally, low polarity solvents and lower temperature are beneficial to the formation of hydrogen bonding interaction in the pre-polymerization complexes and polar solvents are good for ion pair and other strong dipolar interactions. However, it is important to note that the choice of solvent is usually limited. For example, the solubility of the template in solvent must be taken into account in practice and the solvent also has an effect on the structure and porosity of the MIPs. Under these constraints, only a few reagents could be employed as the solvents for MIPs.

The information about the strength and stoichiometry of template-functional monomer interactions helps to understand the nature of the pre-polymerization complex and some experimental studies have been employed to gather the information. Many researches have been proved that the spectroscopic methods, including FT-IR [28], UV-Vis spectrometry [29–31], and nuclear magnetic resonance (NMR) [32–35], are useful to understand the pre-polymerization complex. For instance, by comparing the FT-IR spectra of MIP to the spectra of NIP, whether the molecular imprinting of template occur or not during the preparation can be shown directly [36].

UV-Vis spectrometry is also a well-established method to identify whether the monomer could interact with template and form a stable pre-polymerization complex in a certain solution. Zhang and coworkers [29] have investigated the interaction between MAA and erythromycin by general UV-Vis spectroscopic analysis. The hydroxyl groups and tertiary amine group of the erythromycin can form hydrogen bond and ionic bond with the related monomers, respectively. In this work, MAA was selected as the monomer for the preparation of MIPs. The results proved that the adsorption spectrum of a constant concentration of erythromycin will change with the addition of MAA. If the MAA increased, the adsorption spectrum was found to be red shifted. It is means that new chemical bonds were formed between erythromycin and MAA in the pre-polymerization system. However, FT-IR and UV-Vis spectroscopy are not always sensitive to the changes in the pre-polymerization complex. Application of these two spectroscopic methods is limited in the study of pre-polymerization system.

The NMR technique is the most commonly used for illustrating the interaction type and intensity of the template-monomer complex in non-covalent imprinting systems. The changes of the interactions between template and functional monomer will be reflected in NMR spectrum, because the interaction will lead to local changes in the electronic environment of the parts of the molecules involved. ^{1}H-NMR is currently used in structure analysis of the pre-polymerization complex. In the case of ^{1}H-NMR titrations, chemical shifts in ^{1}H-NMR spectra will change during titrations of the template with the monomer. In the study of Karlsson et al. [37], the bupivacaine MIPs was prepared and a model for the molecular recognition in bupivacaine MIPs has been established based on a series of ^{1}H-NMR titrations.

For simplify the interpretation of the resulting ^{1}H-NMR spectra, acetic acid-d_4 was taken as an analogue for the functional monomer (MAA). Toluene-d_8 and chloroform-d were taken as porogens which are commonly used in the preparation of MIPs. The results shown that the chemical shifts of ^{1}H resonances were changed along with the acetic acid-d_4 concentration increase. The resonance from the amide proton of template (bupivacaine) is primarily influenced by acetic acid-d_4. The downfield shift was observed and it was caused by the formation of hydrogen bonding interactions between template and functional monomer analogue.

Work by Bermejo-Barrera and coworkers [38] details the use of NMR and nuclear overhauser effect (NOE) for elucidating the pre-polymerization complex and the interaction type and intensity of the template-monomer preassembled system was investigated. In this research, cocaine hydrochloride (COCH), MAA, and EDMA was taken as template, functional monomer, and cross-linker, respectively. The hydrogen bonding interactions of MAA–COCH and EDMA–COCH were evaluated by NOE NMR. The results proved that the MAA–COCH interaction formed by the interaction between the protonated amino group of COCH and the carbonyl group of MAA. There is an N–H···O hydrogen bond between template and monomer. The interaction of COCH-EDMA was also studied and it is demonstrated that the COCH-EDMA interaction was formed between the amino group of COCH and the ester carbonyl group of EDMA through N–H···O hydrogen bond. Moreover, the selection of the best template-monomer ratio was also achieved with the aid of NMR.

There have been many researches on the characterization of pre-polymerization complex by ^{1}H-NMR [32, 39–41]. However, this method leads to very complicated ^{1}H-NMR spectra with multiplets. The adjacent chemically non-equivalent H atoms will cause the coupling and the splitting of ^{1}H-NMR spectrum. Dubey and coworkers [42] used ^{31}P{^{1}H}NMR to optimize imprinting conditions of derivative of methylphosphonic acid. Since the chemical shift change for the free template and bound monomer template are in proportional to the strength of interactions, the interactions could be monitored by the ^{31}P{^{1}H} NMR analysis of the template which contains P atom. The advantage of ^{31}P{^{1}H} NMR is that the spectrum only displays one single peak for each P atom within the template. Compared to ^{1}H-NMR, ^{31}P {^{1}H} NMR spectrum can be interpreted more easily. Therefore, ^{31}P{^{1}H} NMR could be an alternative tool for the research the template-monomer interactions.

These researches proved that NMR has been shown to be valuable tool for the preparation of MIPs. Furthermore, some new advanced techniques also applied the characterization of pre-polymerization complex. Quartz crystal microbalance (QCM) technique has been usually used to analyte detection of MIPs in recent years [43–46] and surface plasma resonance (SPR) is especially useful for analyzing the thermodynamics and kinetics of intermolecular interactions between biomolecules [47, 48]. These techniques can help to obtain the information of pre-polymerization complex. However, there are still some limitations to them. For example, a template grafting progression is required before the real measurement. Hence, in order to truthfully characterize the interactions in the pre-organization system, these new advanced techniques need further development in the future. In practice, these instrument analysis techniques mentioned above are frequently used in combination to get as much information of the pre-polymerization complex as possible.

There are many factors that have an effect on the template-monomer interactions. The ratio of template to functional monomer, polymerization temperature and pressure are the important factors and most studies involve the optimization of these factors. A proper proportion of template and functional monomer ensures that high affinity sites can be created in MIPs and nonspecific binding sites can be minimized. Furthermore, the polymerization temperature has been regarded as an important influence on the template-monomer interactions. Since the high temperature is considered to disrupt the interactions, initiator with lower temperatures or photochemical initiation at low temperature is usually employed in practice [49]. It is important to note that free radical polymerization is an exothermic process. Heat management is important for some MIP synthesis methods. For instance, heat convection is severely hampered in the monolithic MIPs or MIPs prepared by bulk polymerization [50]. Meanwhile, because the excess heat can be removed from the MIPs more efficiently, the synthesis methods based on polymerizations in dispersed phases, such as suspension, emulsion, and precipitation polymerization, less affected by exothermic process. The research about effects of pressure on the efficiency of imprinting is relatively small at present. The research by Sellergren et al. [51] shows that the chromatographic capacity factor for an ametryn MIP increases as the pressure of the polymerization increases. However, the studies of the effects of pressure on MIP synthesis are still very limited.

2.4.2 Non-covalent Imprinting with a Single Functional Monomer

The functional monomers play a key role in non-covalent imprinting system. They can form a pre-polymerization complex with the template through functional groups. A suitable functional monomer can strongly interact with the template and it is decisive to generate high affinity binding sites [52, 53]. In most cases, one single functional monomer was employed in the MIPs polymerization. It can be regarded as the simplest and the most widespread method to non-covalent imprinting. However, it does not mean that the character of the pre-polymerization complexes is also simple in this situation. There are some other interactions exist in the pre-polymerization system. For example, the self-association of functional monomer needs to be taken into account.

Since the molecular imprinting technique has been introduced, many functional monomers have been applied in non-covalent imprinting. Some of them have been widely used, while others have only been reported in a few or even one literature. Generally speaking, functional monomers can be classified into acidic, basic, and neutral monomers according to the nature and the acidic monomers, especially carboxylic acid-based monomers, have been considered as the most successful monomers. For example, methacrylic acid (MAA) can be the most popular functional monomer due to its hydrogen bond donor and acceptor characteristics. In fact, carboxylic acid-based monomers can interact with templates in various ways, includes not only H-bond donors and H-bond acceptors but also ion pair formation, weaker dipole-dipole interactions and so on. The study of Zhang et al. [54] explained in detail why MAA has been considered as a "universal" functional monomer for molecular imprinting. It is important to note that the dimerization was once considered as a drawback of MAA in non-covalent imprinting. However, in this research, it was proved that monomer dimerization can actually improve the imprinting efficiency. Because although the number of template binding sites was reduced by monomer dimerization, the number of non-template binding sites was reduced even more. It leads to that the templated sites were increased in percentage terms. Furthermore, MAA have an effect on the structure and morphology of the resulting MIPs. It has been proved that high molar fractions of MAA would result in the large pore size of MIPs which can help improve the binding capacity of the polymers [55].

Furthermore, in a series of studies, Takeuchi and coworker proved that trifluoromethyl acrylic acid (TFMAA) can be a superior functional monomer for non-covalent molecular imprinting. They indicated that more acidic functional monomers could be preferable for imprinting a basic template molecule [56–58]. Until now, many acidic monomers, such as acrylic acid, itaconic acid, and vinylbenzoic acid, have been applied in non-covalent molecular imprinting researches [59–61]. In general, as the most widely used monomers, they should be highly effective and easily available.

Compared with the acidic monomers, there are by now relatively few literatures using basic monomers. Under the premise, the vinyl pyridines are perhaps the most widely used from the basic monomers. It should be noted that vinyl pyridines are acidic in nature. However, in their basic form, they interact strongly with electron deficient aromatic rings, as well as through acid-base interactions and hydrogen bonding. In many studies, vinyl pyridines have proved useful because they often interact strongly with templates. In the research of Kempe and Mosbach [62], (S)-naproxen MIPs was prepared by using 4-vinyl pyridine as functional monomer. It was assumed that 4-vinyl pyridine interacted with the carboxy group in naproxen by ionic interactions. The resulting MIPs were employed as a chiral stationary phase in HPLC and the resolution of racemic naproxen can be achieved on the MIPs stationary phase efficiently. The major drawback of vinyl pyridines is that the strong k–k interactions results quite high levels of nonspecific binding of templates. Furthermore, vinyl imidazole could be a useful monomer. A porous MIP thin-film was fabricated (Fig. 2.7) and used as a microextraction adsorbent for the selective extraction of trace amounts of polycyclic aromatic sulfur heterocycles (PASHs) from seawater [63]. Taken 1-vinylimidazole as the functional monomer, the optimized MIP thin-film was synthesized. The novel MIP thin-films showed good selectivity, excellent binding behavior, and reproducibility for PASHs. Vinyl imidazole monomers also can be used in metal complexation systems, but beyond that it has been

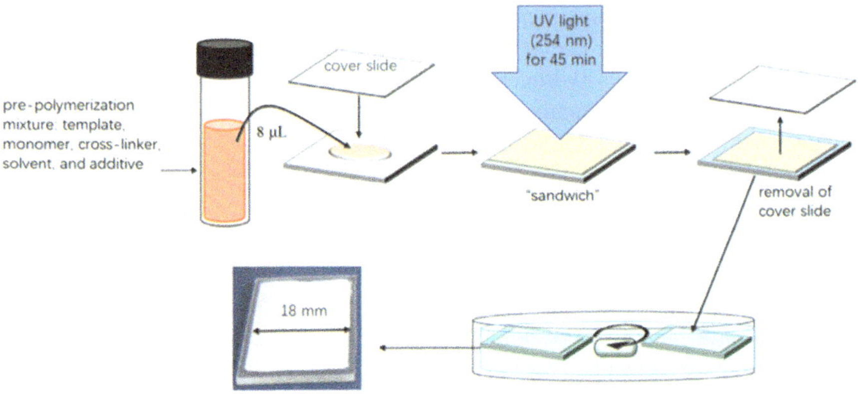

Fig. 2.7 Fabrication process of MIP thin-film on a glass slide. (Reproduced with permission of Ref. [63])

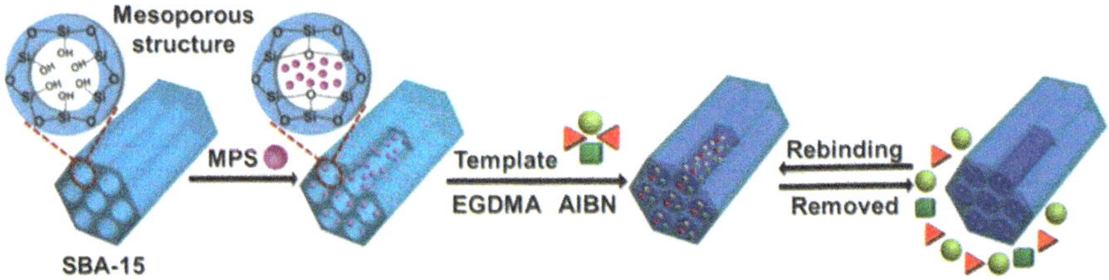

Fig. 2.8 Schematic representation for the preparation of the SBA-15@MT-MIPs. (Reproduced with permission of Ref. [64])

little used. In principle, basic monomers have been effective in some researches, but in general yield less satisfactory results than acidic monomers.

Many neutral monomers have also been widely in molecular imprinting. Acrylamide and its N-alkyl derivatives can be the most successful neutral monomers. In the study of Sun et al. [64], the multi-template MIPs were fabricated using three saponins as multi-template, acrylamide as functional monomer, EGMA as cross-linker, mesoporous silica (SBA-15) as solid support, and ethanol as porogen (Fig. 2.8). The obtained MIPs were reusable and had excellent stability. The materials were employed for the efficient remove impurities and enrichment of the trace level of saponins from plasma samples simultaneously. Furthermore, five different acrylamide-based functional monomers were evaluated by Hayes and coworkers [65]. In this work, myoglobin protein was taken as template and the possible binding interactions between template and the acrylamide-based functional monomers was investigated by using computational technique.

It is worth mentioning that *N*,*O*-bismethacryloyl ethanolamine (NOBE) has aroused extensive interest as a new neutral monomer for molecular imprinting [66]. The NOBE was used to simplify the preparation of MIPs since it combines the functionality of template binding and cross-linking (Fig. 2.9). The corresponding

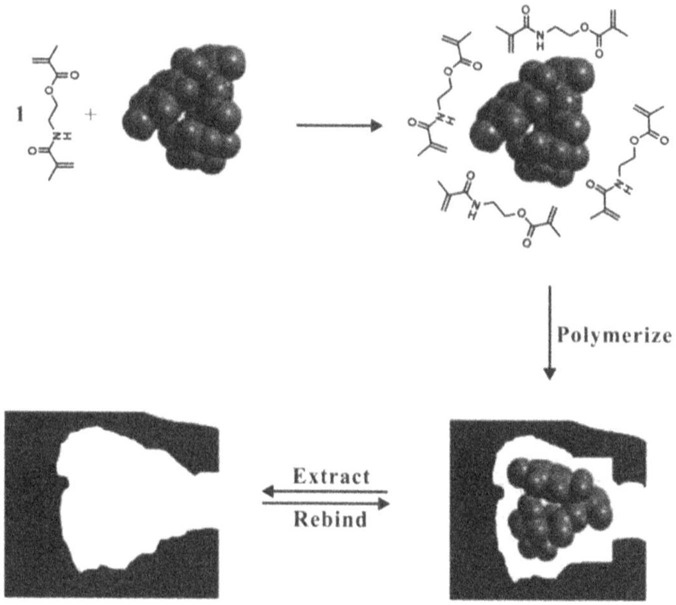

Fig. 2.9 Schematic of the simple OMNiMIP imprinting method with BOC-L-tyrosine as template. (Reproduced with permission of Ref. [69])

studies showed that the higher performance of the one monomer molecularly imprinted polymers (OMNiMIPs) can be achieved compared with the MIPs obtained by the traditional monomer [67–70]. However, this kind of novel monomers for MIPs are still limited by now and more in-depth researches are required.

Some functional monomers have special structures and are hard to categorize according to the criteria mentioned above. For instance, represented by β-cyclodextrins (β-CDs), a series of cyclic oligosaccharides with a hydrophilic exterior and a hydrophobic cavity have been applied as candidate monomers for molecular imprinting [71]. β-CDs can form inclusion complexes with the template through various intermolecular interactions, such as hydrogen bonding, electrostatic interactions, van der Waals forces, and host-guest interactions. In addition, in the presence of a proper cross-linker, the hydroxyl group on β-CDs can take as a polymerization terminal to form a stable polymer matrix. For instance, Liu and coworkers [72] polymerized the aesculin MIPs using β-CD as functional monomers. The study showed that the resulting MIPs have an affinity for template. Moreover, the polymer displayed good controlled release behavior for aesculin.

Miyata and coworkers [73] prepared bisphenol A (BPA) responsive hydrogels with β-CD via MIT. In this work, the β-CDs with acryloyl group were used as ligand (Fig. 2.10). In the pre-polymerization process, a template-functional monomer complex with a sandwich structure, CD-BPA-CD complex, was formed. Moreover,

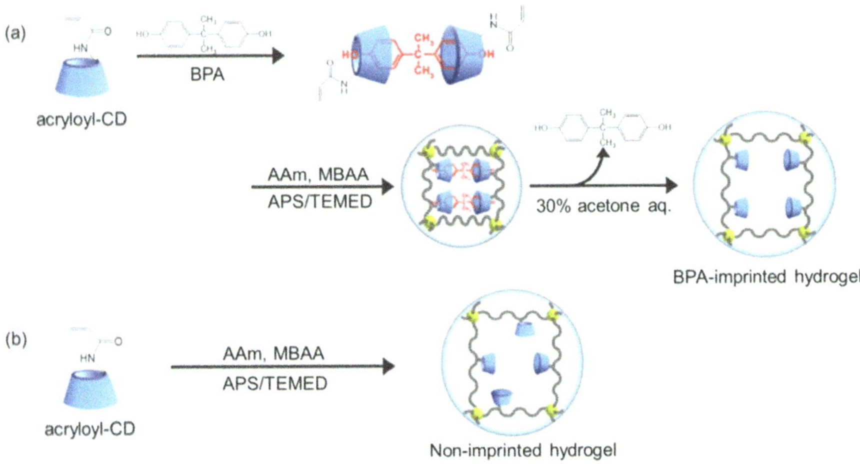

Fig. 2.10 Synthesis of BPA-imprinted (**a**) and non-imprinted hydrogel (**b**). (Reproduced with permission of Ref. [73])

the sandwich-like complex can also act as cross-linker and it makes the apparent cross-linking density of the BPA-imprinted and non-imprinted hydrogels increase.

2.4.3 Imprinting with Combinations of Monomers

The performance of the template-functional monomer interactions is a very important influence to the non-covalent imprinting. The non-covalent interactions can be enhanced by multipoint interactions [74]. Hence, it is very promising to combine the potential specific interaction of different monomers in molecular imprinting. Until now, there are plenty of examples in the literature to demonstrate molecular imprinting by using combinations of two or more functional monomers.

The first application of multi-functional monomer for MIPs was proposed by Mosbach and coworkers [75]. As mentioned above, 2-VP is weakly basic and MAA is acidic. In this study, these two chemically distinct functional monomers were used simultaneously to prepare the MIPs. The resulting MIPs showed improved recognition ability compared with MIPs which were obtained by only one monomer.

Cao and coworkers [76] proposed an efficient strategy for preparation of bifunctional monomers perfluorooctanoic acid (PFOA) MIPs by using 4-vinyl pyridine (4-Vpy) and 2-(trifluoromethyl) acrylic acid (TFMAA) as binary functional monomers (Fig. 2.11). The resulting MIPs was used to specific recognition for PFOA and perfluorooctanesulfonic acid (PFOS) from aqueous solution. The synthesized polymer showed good adsorption capacity and selectivity performance for PFOA and PFOS.

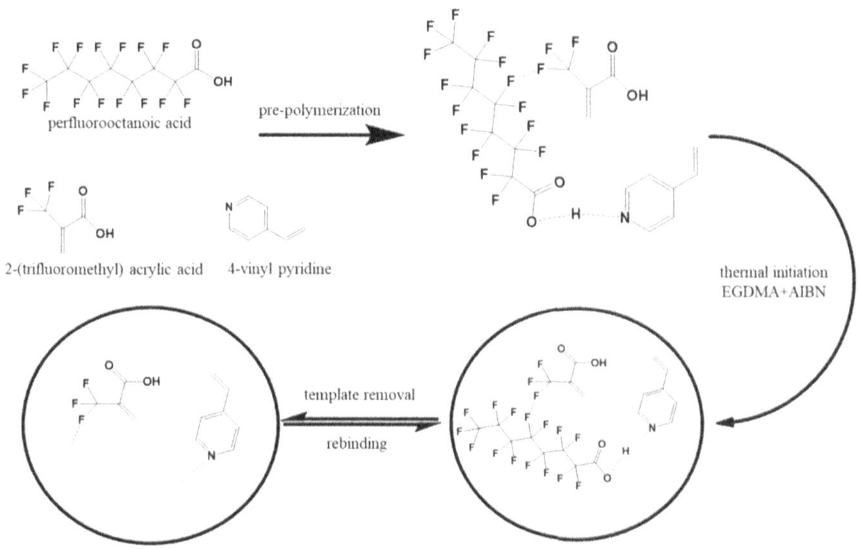

Fig. 2.11 The schematic of the preparation of the binary monomers MIPs. (Reproduced with permission of Ref. [76])

Li et al. [77] prepared MIPs toward norfloxacin (NOR) by using aminopropyltriethoxysilane (3-APTES) and methacryloxypropyltrimethoxysilane (MTEOS) as functional monomers. The MIPs with bifunctional monomer shows a better selectivity for norfloxacin compared with structured analogues and nonstructured analogues. The MIPs was evaluated by various techniques and it was proved that the polymer possessed a good adsorption capacity and an impressive select factor. Of course, there are more applications of MIPs obtained by combinations of functional monomers were proposed by many different research groups [78, 79].

Generally speaking, application of binary functional monomers is more common in the literature. But there are also examples of utilization of multi-functional monomers (more than two) for MIPs. In the study of Haruki and coworker [80], the MIP toward native lysozyme promotes the folding of chemically denatured lysozyme was prepared by using multi-functional monomers (acrylamide, MAA, and 2-(dimethylamino)ethyl methacrylate) as functional monomers. High refolding yield was obtained because of multi-non-covalent interaction between the template and functional monomers.

The application of dual or multiple functional monomers is an effective way to improve the selectivity of MIPs and especially it is a good strategy to imprint macromolecule templates. However, this method should be further improved. For example, the selection of dual/multiple functional monomers for MIP preparation is a hard work due to the complication of pre-polymerization mixture. All kinds of the interactions, such as template-monomer 1, template-monomer 2, and monomer 1-monomer 2, should be considered simultaneously during the optimization.

2.4.4 Custom-Designed Monomers

The common monomers are usually commercially available. However, the performance of these monomers is not good enough for some templates with special structure. The monomers with more hydrogen bonding sites can enhance association constants and the resulting MIPs have more selective recognition sites. For example, a polymerizable bis-urea monomer designed and synthesized by Hall et al. [81]. And the new monomer can interact with the L-glutamyl residue of methotrexate. Barbiturate MIPs were prepared using a new functional monomer by Tanabe et al. [82]. This new monomer, 2,6-bis-acrylamidopyridine, can interact with templates though multiple hydrogen bond. The concave H-bond donor-acceptor-donor configuration of the monomer can match the convex acceptor-donor-acceptor structure of the template. Moreover, the NOBE which is used for the preparation of one monomer MIPs (OMNiMIPs) could be also considered as custom-designed monomers (see Sect. 2.4.2 of this book) [66].

Room temperature ionic liquids (RTILs) could be defined as molten salts with melting points near room temperature. RTILs have been considered as a novel kind of green solvents with interesting and unique property, such as non-volatility, non-volammability, good ion density, and high ionic conductivity etc. RTILs have been increasingly applied in MIT and play multiple roles in MIT. For instance, RTILs are employed as functional monomers and the resulting MIPs have satisfactory recognition ability to template. RTILs could interact with various molecules such as the common organic compounds and biomacromolecules by hydrogen bonding, electrostatic, anion-exchange, π–π interactions and so on [83, 84]. As a "designer solvent," it offers a greater degree of flexibility, which makes RTILs applied for various purposes in MIT. Moreover, many studies have showed that RTIL-based MIPs have excellent performance in aqueous solution.

In the research of Wang's group [85], a new chlorsulfuron MIPs was synthesized by using vinylimidazolium RTIL, 1-vinyl-3-butylimidazolium chloride ([VBIM]Cl), as a unique functional monomer. This kind of new MIPs was prepared by bulk polymerization and showed good selectivity and adsorption/desorption for chlorsulfuron. The binding selectivity of the obtained MIPs was investigated by competitive adsorption using the mixture solution composed of template and related analogues. The novel MIPs revealed good selectivity of 47.2% for template, which was higher than that for the analogues. Furthermore, it is important to notice that low concentration chlorsulfuron could be detected by the proposed MIPs in water solution. Up to now, the studies of the imprinting of normal small organic molecules are very active and plenty of relevant researches have been published. However, fabrication of MIPs for large molecules such as enzyme, polypeptide and proteins has been considered as the toughest challenge in the field of MIPs. Most of the time, this is due to lack of suitable functional monomer for these macromolecules. Fortunately, it is one of effective methods to solve this problem by using RTILs monomers. Thymopentin (TP5) is a biomacromolecule and it is consisted by five amino acids. The purify TP5 is hard to extract from complicated biological samples.

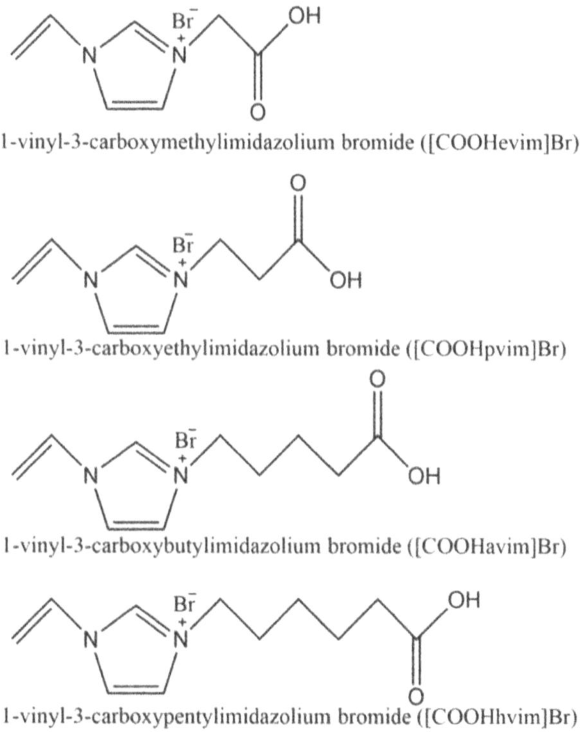

Fig. 2.12 The molecular structure of these four RTIL functional monomers

In a study, a method to prepare TP5 magnetic MIPs by surface-initiated ATRP polymerization was developed [86]. Four candidate RTILs functional monomers, 1-vinyl-3-butyl imidazolium chloride ([VBIM]Cl), 1-vinyl-3-propyl imidazolium chloride ([VPIM]Cl), 1-vinyl-3-ethyl acetate imidazolium chloride and 1-vinyl-3-ethanamide imidazolium chloride, was studied (Fig. 2.12). Molecular dynamics (MD) simulations was employed for computational design of MIPs to find the optimal experimental scheme. In the research, the MD method was employed for investigating the interactions between TP5 and the RTILs monomers.

At last, the TP5 magnetic MIPs were synthesized based on the results of MD simulation. It is proved that the proposed magnetic MIPs prepared with RTILs monomer have excellent specific recognition to TP5.

2.5 Monomer Selection and Optimization Methods

The purpose of molecular imprinting technology is to acquire polymer with high performance. However, it is difficult to predict the performance of any MIPs according to its composition. The optimization of formulation components for MIP preparation is an onerous and time-consuming task. During the optimization process, lots of variables should be considered simultaneously such as the type and the optimum ratio of formulation components for MIPs. The conventional MIPs synthesis and processing is usually based on the prior literature and this leads to a tendency toward sticking with what has worked in the past. The mechanism of imprints formation and recognition are still not entirely clear. Furthermore, variables influencing performance of MIPs were dependent, making it hard to explain how these variables will interact with each other. Up to date, various approaches have been developed to investigate the related mechanism, such as combinatorial approaches, chemometric methods, and molecular modeling approaches. These methods were used to prepare MIPs with high selectivity easily and employed to simplify the preparation of MIPs.

2.5.1 Combinatorial Approaches to Optimization

Use the imprinting process could be affected by many variables, the optimization needs to be achieved by a multivariate strategy. An optimization approach named combinatorial approach was independently developed by two different teams [87, 88]. This method used an automated system to dispense small volumes of different imprinting mixtures into glass vials. It is means that the composition, such as templates monomers, cross-linkers, solvents, and initiators, is systematically varied. After polymerization and extraction of the template, a solution of template is filled into the vials for adsorption measurements. Then, the affinity of the different MIPs could be determined by analyzing the remaining template concentration in the solution. The key to combinatorial approaches is that a sufficient quantity of polymer compositions is synthesized and evaluated. In the related work [88], 96-well microtiter plates were applied to produce MIPs with different compositions. A fluorescence-based screening method was applied to speed up the evaluation of MIPs performance. The optimized MIPs were prepared by using conventional large-scale synthesis and the results can be validated by screening the "MiniMIP" library. The similar approach has been used to develop optimized MIPs for different templates and application formats [89–92].

However, this approach has not been widely used at present. Because in order to make the process automatic, a complex and costly equipment had to be build or bought. Hence, improvement on the method is very necessary. It is means that a simple and fast screening of MIPs can be achieved. For instance, Bruggemann and coworkers [93] proposed an approach based on ultrafiltration membrane modules. A MIPs membrane was prepared and used for screening procedures. In this approach, the affinity of each MIPs membrane toward the template is studied by pumping a defined amount of the template solution through the MIPs membrane. By measuring the concentration of the template in the permeate solution, the adsorbed amount which is related to the control polymer membrane can be obtained. The results proved that the proposed method can be used for finding optimized template-cross-linker ratios in different porogens by screening MIPs of various compositions.

2.5.2 Chemometric Methods

Chemometrics could be considered as a large class of approaches, which applies the approaches of mathematics, statistics, and computer technology. In practice, chemometric methods are often used to the optimization of system parameters with a significant impact on the synthesis of MIP and corresponding physical and chemical performance [94, 95]. Conventional approaches for MIP optimization only generate limited information although a large amount of experimental research has been made. By contrast, the relevant variables could be optimized systematically and simultaneously with the help of chemometrics in experiment design. Obviously, compared with the conventional univariate experimentations, the experiment designed by chemometrics require less measurements. Furthermore, chemometrics could also be applied to reveal the interaction between the variables. Hence, these advantages of chemometrics makes the optimization process of MIP preparation much efficient and simpler [96].

In the research of Li and coworkers [97], a rational design of photonic MIP films with the aid of chemometrics was proposed. The response surface methodology (RSM) based on central composite design (CCD) was used to the design of the novel photonic materials (Fig. 2.13). RSM was applied to explore the interactive influence of different fabrication parameters and the key parameters related to the sensing performance of the photonic MIP were confirmed with the aid of chemometric method. As a result, the dominant factor (cross-linker) for the preparation of photonic MIPs can be determined and the optimum ratio of monomer, cross-linker and solvent can also be confirmed which made the production of materials the best template molecular recognition abilities. This research proved that the chemometrics can be applied to optimize the factors which were associated with the preparation of photonic MIPs and the optimum formulations of materials can be obtained with fewer experiments.

For the optimization approaches of conventional MIP, univariate experiments must be carried out, which is laborious, time-consuming, and uneconomic. Davies et al. [94] proposed an effective chemometric method for the optimization of MIPs. In the research, a rational design method named three-level full factorial design was applied for the preparation and optimization of sulfonamides MIP. With the aid of the chemometrics, the optimum ratio of template: monomer: cross-linker for the sulfonamides MIP can be predicted exactly. In another work [98], rational design was used to select cross-linker. For the preparation of zidovudine (AZT) MIP, two candidate cross-linkers, divinylbenzene (DVB) and trimethylolpropane trimethacrylate (TRIM), were investigated. The results obtained by adsorption experiments and molecular modeling showed that the DVB cross-linker was more suitable for AZT MIPs. Furthermore, the results obtained by the static adsorption experiments showed that the DVB-based AZT MIPs had the highest imprinting factor. It is means that the optimal cross-linker could be selected based on the strength of the template-cross-linker interaction. In the study of Muhammad et al. [99], a water-compatible MIPs was prepared. To simplify the experimental process,

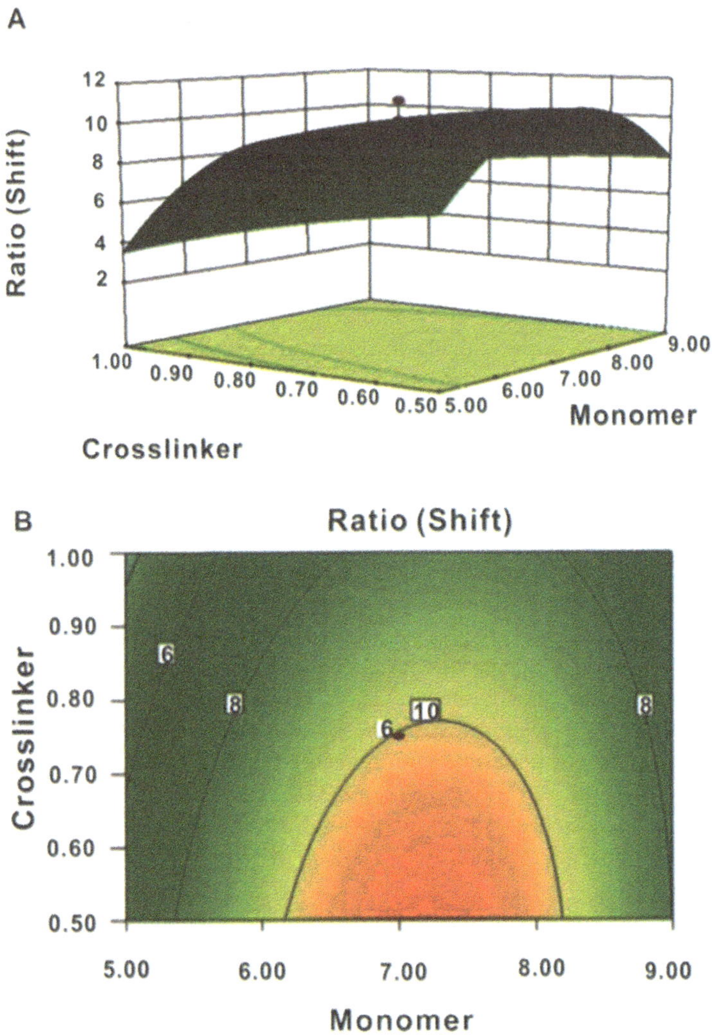

Fig. 2.13 Response surface plot (**a**) and contour plot (**b**) of the combined effects of the monomer and cross-linker on ratio (shift). (Reproduced with permission of Ref. [97])

a new rational design approach based on screening library of non-imprinted polymers (NIPs) was developed. The organized NIP library contains 18 cross-linked co-polymers. These co-polymers were obtained by the monomers which are normal to MIP. In this work, 4-vinylpyridine (4-VP) was selected as the most suitable monomer for preparing amiodarone MIPs by using the proposed method. This research also indicated that a good correlation of the screening tests and modeling of template-monomer interactions can be obtained by the computational approach.

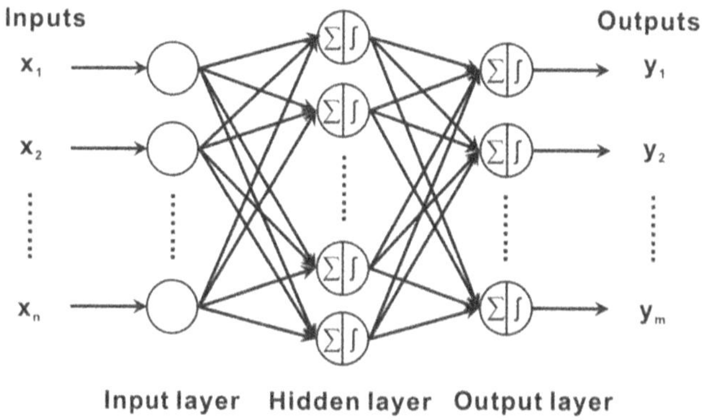

Fig. 2.14 Schematic diagram of a three-layer feed-forward back-propagation neural network used in Ref. [100]. Circles represent neurons and the connection between neurons represents weights. The summation and sigmoid symbol represents summation and sigmoid transfer function, respectively. (Reproduced with permission of Ref. [100])

The results proved that the proposed method is a potential common computational and combinatorial approach for the preparation of MIPs.

Artificial neural networks (ANNs) are a kind of bionic algorithms by imitating biological neural networks and it has found an increasingly wide utilization in many fields. ANN could obtain the best approximation of the practical problems by large learning and training. In the research of Prachayasittikul and coworkers [100], ANN has been brought into the molecular imprinting technique. Briefly, the computed molecular descriptors of template, functional monomer and mobile phase descriptors were chosen according to the magnitude of imprinting factors (IF) of MIP. Then, the IF was calculated by ANN and the results showed that ANN can be considered as a powerful tool for predicting the feasibility of potential template-functional monomer complex before the practical experiments (Fig. 2.14). The unique estimation ability of ANN provided insights on the feasibility of the interaction between template and monomer.

The experiment design has also been applied to optimize the reaction parameters' impact on the performance of template recognition for MIPs preparation.

Kempe and Kempe [101] used multivariate data analysis and the statistical experimental design to optimize the parameters in the model. In their research, MIP bead libraries were built by using the statistical experimental design. The amounts of MAA, TRIM and acetonitrile were taken as the parameters to optimize. The ratio of the amount of free template to the amount binding in the pre-polymerization namely partition coefficient was taken as the response information. Meanwhile, experimental design was achieved by a composite face-centered (CCF) quadratic model. The data were processed by the multiple linear regression (MLR) and the corresponding results were used to demonstrate the prediction and the goodness of fit of this approach.

MIPs usually have a vast range of binding affinities since MIPs have hydrophobic surface and hydrophilic surface simultaneously. Hence, MIPs have specific and nonspecific binding sites. The balance between electrostatic interactions and hydrophobic would influence the type of binding sites on MIP surface. As a result, it is difficult to discriminate the function from binding data. Moreover, the study on MIPs binding ability with the template is also a very important task for researchers. Nicholls and coworkers [102] used chemometrics to describe and predict the binding extent between template and MIPs for the first time. In this chapter, equilibrium binding study was applied to study the bonding degree of template (bupivacaine) with MIPs and reference polymers. The measuring data is processed by the partial least-squares regression (PLSR) and the corresponding third-degree equations can be obtained. As a result, the corresponding chemometric models were built for describing template binding in the chosen system. The research shows that these models have good correlation and predictive ability. Meanwhile, this study proved that temperature and dielectric constant could be used to describe binding. The results showed that temperature has little impact on the binding ability in the nonpolar and aprotic solution. In contrast, hydrophobic interactions and temperature have important influence on binding in polar solution.

Principal component analysis (PCA) is a kind of effective data compression method. It can compress multidimensional data linearly into lower dimensions with minimal loss of information. In order to identify the most important factor for the rebinding among the template and the corresponding MIPs, Nicholls and coworkers [103] also investigated physical properties of the media on MIP performance by processing rebinding data with PCA. The results proved that polarity of solvent and the dielectric constant (D) have the greatest effects on the binding. At the same time, the chemometrics methods represented by PCA are powerful tools for exploring the true relationship between the corresponding factors and the rebinding ability of MIPs. The mathematical models with good performance for the binding process can be built by chemometrics.

Moreover, the template-functional monomer complex which was formed in the pre-polymerization process has a significant effect on the recognition and morphology of MIPs. PCA was also used to study the effect of template complexation in pre-polymerization mixtures on MIP recognition and morphology [104]. In this study, a series of MIPs which were prepared by changing the ratio of template (bupivacaine): functional monomer (MAA or methyl methacrylate): cross-linker (EGDMA) were obtained. The corresponding MIPs were characterized by swelling studies, gas sorption measurements, and radioligand equilibrium binding experiments. Molecular dynamics (MD) simulation was employed to extract information from pre-polymerization complexes. PCA was applied to process information which was obtained from all-component MD simulation trajectories of a series of MIP pre-polymerization complexes. Furthermore, the data describing the surface characteristics and rebinding behaviors of similar synthesized polymers was also processed by PCA. The above results proved that PCA can be used to reveal relationships between MD-derived descriptions of events in the pre-polymerization complex, recognition performance and morphologies of MIPs.

2.5.3 Molecular Modeling Approaches

Up to now, it turns out that computational calculation is an effective way to reduce experimental trials for preparation of MIPs. Based on practice experience, using efficient molecular modeling to assist the optimization of the imprinting protocol is a very good choice. Molecular dynamics (MD) simulation is considered as a classical computational technique and has been used to simulate and predict the best imprinting protocol. Similarly, density functional theory (DFT) is also usually employed for computational design of the scheme for preparation of MIPs. These molecular modeling approaches have been frequently applied for screening the most suitable functional monomer to a target molecule.

In 2001, Piletsky et al. [105] first proposed a MD-based method for the preparation of ephedrine MIPs. In this method, an artificial library of functional monomers was built at first. Then, this library was screened by using the software of molecular modeling in order to select an optimal monomer for template. An algorithm named LEAPFROG is the core of this approach and it was employed to screen the functional monomers library for study the potential mechanisms between template and functional monomers. The score of empirical binding energy can be obtained after 30,000 runs. This score was evaluated and four functional monomers with the best binding score were selected in this study. The computed results indicated these monomers have the best ability of forming the strongest complexes with the template and they were used for the preparation of MIPs. This research implied that the developed computational approach can be considered as an alternative way for the rational design of MIPs. Moreover, by comparing the corresponding binding scores of different monomers, the specificity and affinity of MIPs can be predicted.

Similar studies were developed by Wei et al. [106]. MD simulation was employed to investigate the template-monomer interaction for optimize the polymerization factors and study the mechanism of recognition in MIPs. By the purposed method, the best monomers for 17-estradiol (BE2) MIPs were selected from a nine functional monomer library by comparing the strength of hydrogen bonding. According to the simulations results, 2-(diethylamino) ethyl methacrylate (DEAEMA), methacrylamide and MAA were recognized as the most suitable monomers for template. The results showed that they have strong affinity to 17-estradiol. The theoretical prediction results were in agreement with previous studies which implied the same functional monomers for the preparation of MIPs. Furthermore, the influence of different temperature and cross-linkers on the performance of the MIPs was studied by Nicholls and coworkers [107]. On the basis of the cross-linkers were different, the template-functional monomer interactions were modeled by MD simulation. The results of simulation explored the most relevant sites of templates which interacted with monomer and it also illustrated the interactions among template, functional monomer, and cross-linker. Furthermore, the interactions between functional monomer and cross-linker were also studied.

In addition to the MIPs design, MIP behavior also can be predicted and explained by the MD methods. For instance, Henthorn and Peppas [108] investigated the

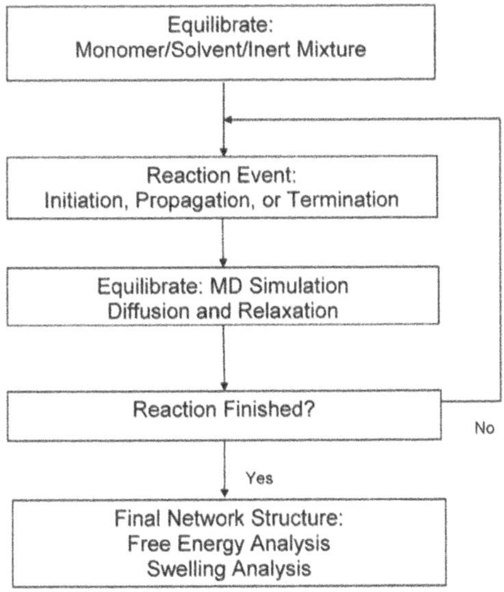

Fig. 2.15 Schematic diagram of kinetic gelation reaction. The reaction is started after the initial equilibration. Radicals are created at every reaction step, allowed to propagate or terminated. Reaction continues until a final conversion is achieved at which time the final network structure is given. (Reproduced with permission of Ref. [108])

densely cross-linked polymeric networks formation mechanism by using a technique named all-atom kinetic gelation which was used to simulate the synthesis system. From beginning to end of this simulation, the interaction and position of all atoms was tracked by an off-lattice method in this work (Fig. 2.15). By using the proposed method, the interactions between the reacting monomers and the recognition mechanism of the polymeric network formation were exploited. It also proved that the network structure of polymers was also effected by templates in this case.

In another study, the interactions between two monomers in an aqueous pre-polymerisation system were studied by using molecular modeling method [109]. In this case, 2, 4-dichlorophenoxyacetic acid (2, 4-D) and 4-VP were taken as the study subjects and the results showed proposed method provided an alternative way to illustrate how MIPs with good performance are prepared in an aqueous solvent.

Another popular quantum method, DFT, also has usually been applied for optimization of geometries of the template-monomers complexes and calculation of interaction energy. It means that DFT can be applied in the design of MIPs. In the study of Ahmadi et al. [110], a methadone MIP was prepared with the help of DFT. According to the calculated results of interaction energy obtained with DFT, MAA was confirmed as the most suitable monomer. This research suggests that the performance of DFT designed MIPs was superior to the MD designed MIPs because the non-covalent template-monomer interactions was represented by DFT. Similarly, with the aid of DFT, MIPs were prepared by Ahmadi et al. [111, 112]. This polymer was applied for the analysis of drugs from plasma. The most suitable monomer for those different drugs were confirmed by the computational methods from DFT. Up to now, DFT is often used to study the interaction of template-monomer complexes to identify the most stable structure [113, 114]. It should be noted that the solvent is also an important variable. The selection of solvents is another important task for the rational designed MIPs. In computational chemistry, polarizable continuum model (PCM) is also often used to investigate the influence of solvent. PCM is a type of continuum model and has been used in lots of researches [115, 116]. In practice, the computational approaches, such as DFT, MD and PCM, were frequently applied simultaneously in a lot of reported literatures [117–120].

References

1. Wulff G, Biffis A (2001) Molecular imprinting with covalent or stoichiometric non-covalent interactions. Elsevier, Amsterdam
2. Wulff G, Dederichs W, Grotstollen R, Jupe C (1982) On the chemistry of binding sites, Part II. Elsevier, Amsterdam
3. Wulff G, Lauer M, Bohnke H (1984) Rapid proton transfer as cause of an unusally large neighboring group effect. Angew Chem Int Ed Engl 23:741–742
4. Wulff G (1995) Molecular imprinting in cross-linked materials with the aid of molecular templates-a way towards artificial antibodies. Angew Chem Int Ed Engl 34:1812–1832
5. Wulff G, Schauhoff S (1991) Racemic resolution of free sugars with macroporous polymers prepared by molecular imprinting. Selectivity dependence on the arrangement of functional groups versus spatial requirements. J Org Chem 56:395–400
6. Alexander C, Smith CR, Whitcombe MJ, Vulfson EN (1999) Imprinted polymers as protecting groups for regioselective modification of polyfunctional substrates. J Am Chem Soc 121:6640–6651
7. Wulff G, Vietmeier J (1989) Enzyme-analogue built polymers, 26. Enantioselective synthesis of amino acids using polymers possessing chiral cavities obtained by an imprinting procedure with template molecules. Makromol Chem 190:1727–1735
8. Wulff G, Best W, Akelah A (1984) Enzyme-analogue built polymers, 17. Investigations on the racemic resolution of amino acids. React Polym Ion Exch Sorbents 2:167–174
9. Wulff G, Heide B, Helfmeier G (1986) Molecular recognition through the exact placement of functional groups on rigid matrices via a template approach. J Am Chem Soc 108:1089–1091
10. Shea KJ (1994) Molecular imprinting of synthetic network polymers: the de novo synthesis of macromolecular binding and catalytic sites. Trends Polym Sci 2:166–173
11. Shea KJ, Sasaki DY (1989) On the control of microenvironment shape of functionalized network polymers prepared by template polymerization. J Am Chem Soc 111:3442–3444
12. Shea KJ, Sasaki DY (1991) An analysis of small-molecule binding to functionalized synthetic polymers by 13C CP=MAS NMR and FT-IR spectroscopy. J Am Chem Soc 113:4109–4120
13. Joshi VP, Karode SK, Kulkarni MG, Mashelkar RA (1998) Novel separation strategies based on molecularly imprinted adsorbents. Chem Eng Sci 53:2271–2284
14. Whitcombe MJ, Vulfson E (1995) A new method for the introduction of recognition site functionality into polymers prepared by molecular imprinting: synthesis and characterization of polymeric receptors for cholesterol. J Am Chem Soc 117:7105–7111
15. Cheong SH, McNiven S, Rachkov AE, Levi R, Yano K, Karube I (1997) Testosterone receptor binding mimic constructed using molecular imprinting. Macromolecules 30:1317–1322
16. Mayes AG, Whitcombe MJ (2005) Synthetic strategies for the generation of molecularly imprinted organic polymers. Adv Drug Deliv Rev 57:1742–1778
17. Qi PP, Wang JC, Wang LD, Li Y, Jin J, Su F, Tian YZ, Chen JP (2010) Molecularly imprinted polymers synthesized via semi-covalent imprinting with sacrificial spacer for imprinting phenols. Polymer 51:5417–5423

18. Ki CD, Oh C, Oh SG, Chang JY (2002) The use of a thermally reversible bond for molecular imprinting of silica spheres. J Am Chem Soc 124:14838–14839
19. Hwang CC, Lee WC (2002) Chromatographic characteristics of cholesterol-imprinted polymers prepared by covalent and non-covalent imprinting methods. J Chromatogr A 962:69–78
20. Kirsch N, Alexander C, Lubke M, Whitcombe MJ, Vulfson EN (2000) Enhancement of selectivity of imprinted polymers via post-imprinting modification of recognition sites. Polymer 41:5583–5590
21. Lubke M, Whitcombe MJ, Vulfson EN (1998) A novel approach to the molecular imprinting of polychlorinated aromatic compounds. J Am Chem Soc 120:13342–13348
22. Petcu M, Cooney J, Cook C, Lauren D, Schaare P, Holland P (2001) Molecular imprinting of a small substituted phenol of biological importance. Anal Chim Acta 435:49–55
23. Petcu M, Schaare PN, Cook CJ (2004) Propofol-imprinted membranes with potential applications in biosensors. Anal Chim Acta 504:73–79
24. Patel A, Fouace S, Steinke JHG (2004) Novel stereoselective molecularly imprinted polymers via ring-opening metathesis polymerization. Anal Chim Acta 504:53–62
25. Patel A, Fouace S, Steinke JHG (2003) Enantioselective molecularly imprinted polymers via ring-opening metathesis polymerization. Chem Commun 1:88–89
26. Graham AL, Carlson CA, Edmiston PL (2002) Development and characterization of molecularly imprinted sol-gel materials for the selective detection of DDT. Anal Chem 74:458–467
27. Caro E, Masque N, Marce RM, Borrull F, Cormack PAG, Sherrington DC (2002) Non-covalent and semi-covalent molecularly imprinted polymers for selective on-line solid-phase extraction of 4-nitrophenol from water samples. J Chromatogr A. 963:169–178
28. Zhang ZH, Yang X, Zhang HB, Zhang ML, Luo LJ, Hu YF, Yao SZ (2011) Novel molecularly imprinted polymers based on multi-walled carbon nanotubes with binary functional monomer for the solid-phase extraction of erythromycin from chicken muscle. J Chromatogr B 879:1617–1624
29. Song SQ, Wu AB, Shi XZ, Li RX, Lin ZX, Zhang DB (2008) Development and application of molecularly imprinted polymers as solid-phase sorbents for erythromycin extraction. Anal Bioanal Chem 390:2141–2150
30. Andersson HS, Nicholls IA (1997) Spectroscopic evaluation of molecular imprinting polymerization systems. Bioorg Chem 25:203–211
31. Lu Y, Li CX, Zhang HS, Liu XH (2003) Study on the mechanism of chiral recognition with molecularly imprinted polymers. Anal Chim Acta 489:33–43
32. Sellergren B, Lepisto M, Mosbach K (1988) Highly enantioselective and substrate-selective polymers obtained by molecular imprinting utilizing noncovalent interactions-NMR and chromatographic studies on the nature of recognition. J Am Chem Soc 110:5853–5860
33. Whitcombe MJ, Martin L, Vulfson EN (1998) Predicting the selectivity of imprinted polymers. Chromatographia 47:457–464
34. Matsui J, Miyoshi Y, Doblhoff-Dier O, Takeuchi T (1995) A molecularly imprinted synthetic polymer receptor selective for atrazine. Anal Chem 67:4404–4408
35. Idziak L, Benrebouh A, Deschamps F (2001) Simple NMR experiments as a means to predict the performance of an anti-17 alpha-ethynylestradiol molecularly imprinted polymer. Anal Chim Acta 435:137–140
36. Li S, Ge Y, Tiwari A, Wang S, Turner APF, Piletsky SA (2011) 'On/off'-switchable catalysis by a smart enzyme-like imprinted polymer. J Catalysis 278:173–180
37. Karlsson JG, Karlsson B, Andersson LI, Nicholls IA (2004) The roles of template complexation and ligand binding conditions on recognition in bupivacaine molecularly imprinted polymers. Analyst 129:456–462
38. Sanchez-Gonzalez J, Pena-Gallego A, Sanmartin J, Bermejo AM, Bermejo-Barrera P, Moreda-Pineiro A (2019) NMR spectroscopy for assessing cocaine-functional monomer interactions when preparing molecularly imprinted polymers. Microchem J 147:813–817
39. Manesiotis P, Hall AJ, Sellergren B (2005) Improved imide receptors by imprinting using pyrimidine-based fluorescent reporter monomers. J Org Chem 70:2729–2738

40. O'Mahony J, Molinelli A, Nolan K, Smyth MR, Mizaikoff B (2006) Anatomy of a successful imprint: analysing the recognition mechanisms of a molecularly imprinted polymer for quercetin. Biosens Bioelectron 21:1383–1392
41. Svenson J, Karlsson JG, Nicholls IA (2004) ^{1}H nuclear magnetic resonance study of the molecular imprinting of (−)-nicotine: template self-association, a molecular basis for cooperative ligand binding. J Chromatogr A 1024:39–44
42. Roy KS, Mazumder A, Goud DR, Dubey DK (2018) A simplistic designing of molecularly imprinted polymers for derivative of nerve agents marker using $^{31}P\{^{1}H\}$NMR. Eur Polym J 98:105–115
43. Zhao XL, He Y, Wang YN, Wang S, Wang JP (2020) Hollow molecularly imprinted polymer based quartz crystal microbalance sensor for rapid detection of methimazole in food samples. Food Chem 309:125787
44. Dayal H, Ng WY, Lin XH, Li SFY (2019) Development of a hydrophilic molecularly imprinted polymer for the detection of hydrophilic targets using quartz crystal microbalance. Sensor Actuat B Chem 300:127044
45. Reimhult K, Yoshimatsu K, Risveden K, Chen S, Ye L, Krozer A (2008) Characterization of QCM sensor surfaces coated with molecularly imprinted nanoparticles. Biosens Bioelectron 23:1908–1914
46. Alenus J, Ethirajan A, Horemans F, Weustenraed A, Csipai P, Gruber J, Peeters M, Cleij TJ, Wagner P (2013) Molecularly imprinted polymers as synthetic receptors for the QCM-D-based detection of L-nicotine in diluted saliva and urine samples. Anal Bioanal Chem 405:6479–6487
47. Matsui J, Takayose M, Akamatsu K, Nawafune H, Tamaki K, Sugimoto N (2009) Molecularly imprinted nanocomposites for highly sensitive SPR detection of a nonaqueous atrazine sample. Analyst 134:80–86
48. Bompart M, De Wilde Y, Haupt K (2010) Chemical nanosensors based on composite molecularly imprinted polymer particles and surface-enhanced Raman scattering. Adv Mater 22:2343–2348
49. O'Shannessy DJ, Ekberg B, Mosbach K (1989) Molecular imprinting of amino-acid derivatives at low-temperature (0 °C) using photolytic homolysis of azobisnitriles. Anal Biochem 177:144–149
50. Piletsky SA, Piletska EV, Karim K, Freebairn KW, Legge CH, Turner APF (2002) Polymer cookery: influence of polymerization conditions on the performance of molecularly imprinted polymers. Macromolecules 35:7499–7504
51. Sellergren B, Dauwe C, Schneider T (1997) Pressure-induced binding sites in molecularly imprinted network polymers. Macromolecules 30:2454–2459
52. Kriz D, Ramstrom O, Mosbach K (1997) Peer reviewed: molecular imprinting: new possibilities for sensor technology. Anal Chem 69:345–349
53. Matsui J, Miyoshi Y, Takeuchi T (1995) Fluoro-functionalized molecularly imprinted polymers selective for herbicides. Chem Lett 11:1007–1008
54. Zhang Y, Song D, Lanni LM, Shimizu KD (2010) Importance of functional monomer dimerization in the molecular imprinting process. Macromolecules 43:6284–6294
55. Golker K, Karlsson BRC, Olsson GD, Rosengren AM, Nicholls IA (2013) Influence of composition and morphology on template recognition in molecularly imprinted polymers. Macromolecules 46:1408–1414
56. Matsui J, Doblhoff-Dier O, Takeuchi T (1997) 2-(Trifluoromethyl)acrylic acid: a novel functional monomer in noncovalent molecular imprinting. Anal Chim Acta 343:1–4
57. Matsui J, Takeuchi T (1997) A molecularly imprinted polymer rod as nicotine selective affinity media prepared with 2-(trifluoromethyl)acrylic acid. Anal Commun 34:199–200
58. Matsui J, Nicholls IA, Takeuchi T (1998) Molecular recognition in cinchona alkaloid molecular imprinted polymer rods. Anal Chim Acta 365:89–93
59. Sellergren B, Ekberg B, Mosbach K (1985) Molecular imprinting of amino-acid derivatives in macroporous polymers-demonstration of substrate-selectivity and enantio-selectivity by

chromatographic resolution of racemic mixtures of aminoacid derivatives. J Chromatogr A 347:1–10
60. Suedee R, Songkram C, Petmoreekul A, Sangkunakup S, Sankasa S, Kongyarit N (1999) Direct enantioseparation of adrenergic drugs via thin-layer chromatography using molecularly imprinted polymers. J Pharm Biomed Anal 19:519–527
61. Andersson L, Sellergren B, Mosbach K (1984) Imprinting of amino-acid derivatives in macroporous polymers. Tetrahedron Lett 25:5211–5214
62. Kempe M, Mosbach K (1994) Direct resolution of naproxen on a noncovalently molecularly imprinted chiral stationary-phase. J Chromatogr A 664:276–279
63. Hijazi HY, Bottaro CS (2020) Molecularly imprinted polymer thin-film as a micro-extraction adsorbent for selective determination of trace concentrations of polycyclic aromatic sulfur heterocycles in seawater. J Chromatogr A 1617:460824
64. Sun C, Wang J, Huang J, Yao D, Wang CZ, Zhang L, Hou S, Chen L, Yuan CS (2017) The multi-template molecularly imprinted polymer based on SBA-15 for selective separation and determination of panax notoginseng saponins simultaneously in biological samples. Polymers 9:653
65. Sullivan MV, Dennison SR, Archontis G, Reddy SM, Hayes JM (2019) Toward rational design of selective molecularly imprinted polymers (MIPs) for proteins: computational and experimental studies of acrylamide based polymers for myoglobin. J Phys Chem B 123:5432–5443
66. Sibrian-Vazquez M, Spivak DA (2004) Molecular imprinting made easy. J Am Chem Soc 126:7827–7833
67. Sibrian-Vazquez M, Spivak DA (2003) Improving the strategy and performance of molecularly imprinted polymers using cross-linking functional monomers. J Org Chem 68:9604–9611
68. Menga AC, LeJeunea J, Spivaka DA (2009) Multi-analyte imprinting capability of OMNiMIPs versus traditional molecularly imprinted polymers. J Mol Recognit 22:121–128
69. LeJeune J, Spivak DA (2007) Chiral effects of alkyl-substituted derivatives of N, O-bismethacryloyl ethanolamine on the performance of one monomer molecularly imprinted polymers (OMNiMIPs). Anal Bioanal Chem 389:433–440
70. Wei ZH, Wu X, Zhang B, Li R, Huang YP, Liu ZS (2011) Coatings of one monomer molecularly imprinted polymers for open tubular capillary electrochromatography. J Chromatogr A 1218:6498–6504
71. Lay S, Ni XF, Yu HN, Shen SR (2016) State-of-the-art applications of cyclodextrins as functional monomers in molecular imprinting techniques: a review. J Sep Sci 39:2321–2331
72. Zhao L, Wang XL, Ma L, Shang PP, Huang YP, Liu ZS (2019) Improving affinity of β-cyclodextrin-based molecularly imprinted polymer using room temperature ionic liquid. Eur Polym J 116:275–282
73. Kawamura A, Kiguchi T, Nishihata T, Uragami T, Miyata T (2014) Target molecule-responsive hydrogels designed via molecular imprinting using bisphenol A as a template. Chem Commun 50:11101–11103
74. Athikomrattanakul U, Gajovic-Eichelmann N, Scheller FW (2011) Thermometric sensing of nitrofurantoin by noncovalently imprinted polymers containing two complementary functional monomers. Anal Chem 83:7704–7711
75. Ramstroem O, Andersson LI, Mosbach K (1993) Recognition sites incorporating both pyridinyl and carboxy functionalities prepared by molecular imprinting. J Org Chem 58:7562–7564
76. Cao FM, Wang L, Tian Y, Wu FC, Deng CB, Guo QW, Sun HW, Lu SY (2017) Synthesis and evaluation of molecularly imprinted polymers with binary functional monomers for the selective removal of perfluorooctanesulfonic acid and perfluorooctanoic acid. J Chromatogr A 1516:42–53

77. Li GY, Zha J, Niu MC, Hu F, Hui XH, Tang TY, Fizir M, He H (2018) Bifunctional monomer molecularly imprinted sol-gel polymers based on the surface of magnetic halloysite nanotubes as an effective extraction approach for norfloxacin. Appl Clay Sci 162:409–417
78. Hoai NT, Yoo DK, Kim D (2010) Batch and column separation characteristics of copper-imprinted porous polymer micro-beads synthesized by a direct imprinting method. J Hazard Mater 173:462–467
79. Chen X, Zhang ZH, Yang XX, Liu YN, Li JX, Peng MJ, Yao SZ (2012) Novel molecularly imprinted polymers based on multiwalled carbon nanotubes with bifunctional monomers for solid-phase extraction of rhein from the root of kiwi fruit. J Sep Sci 35:2414–2421
80. Haruki M, Konnai Y, Shimada A, Takeuchi H (2007) Molecularly imprinted polymer-assisted refolding of lysozyme. Biotechnol Prog 23:1254–1257
81. Hall AJ, Achilli L, Manesiotis P, Quaglia M, De Lorenzi E, Sellergren B (2003) A substructure approach toward polymeric receptors targeting dihydrofolate reductase inhibitors: 2. Molecularly imprinted polymers against Z-l-glutamic acid showing affinity for larger molecules. J Org Chem 68:9132–9135
82. Tanabe K, Takeuchi T, Matsui J, Ikebukuro K, Yano K, Karube I (1995) Recognition of barbiturates in molecularly imprinted copolymers using multiple hydrogen-bonding. J Chem Soc Chem Commun 22:2303–2304
83. Yuan SF, Deng QL, Fang GZ, Pan MF, Zhai XR, Wang S (2012) A novel ionic liquid polymer material with high binding capacity for proteins. J Mater Chem 22:3965–3972
84. Desai RK, Streefland M, Wijffels RH, Eppink MHM (2014) Extraction and stability of selected proteins in ionic liquid based aqueous two phase systems. Green Chem 16:2670–2679
85. Guo L, Deng QL, Fang GZ, Gao W, Wang S (2011) Preparation and evaluation of molecularly imprinted ionic liquids polymer as sorbent for on-line solid-phase extraction of chlorsulfuron in environmental water samples. J Chromatogr A 1218:6271–6277
86. Wang CL, Hu XL, Guan P, Qian LW, Wu DF, Li J (2014) Thymopentin magnetic molecularly imprinted polymers with room temperature ionic liquids as a functional monomer by surface-initiated ATRP. Int J Polym Anal Charact 19:70–82
87. Lanza F, Sellergren B (1999) Method for synthesis and screening of large groups of molecularly imprinted polymers. Anal Chem 71:2092–2096
88. Takeuchi T, Seko A, Matsui J, Mukawa T (2001) Molecularly imprinted polymer library on a microtiter plate. Highthroughput synthesis and assessment of cinchona alkaloidimprinted polymers. Instrum Sci Technol 29:1–9
89. Quaglia M, Chenon K, Hall AJ, De Lorenzi E, Sellergren B (2001) Target analogue imprinted polymers with affinity for folic acid and related compounds. J Am Chem Soc 123:2146–2154
90. Ferrer I, Lanza F, Tolokan A, Horvath V, Sellergren B, Horvai G, Barcelo D (2000) Selective trace enrichment of chlorotriazine pesticides from natural waters and sediment samples using terbuthylazine molecularly imprinted polymers. Anal Chem 72:3934–3941
91. Dirion B, Lanza F, Sellergren B, Chassaing C, Venn R, Berggren C (2002) Selective solid phase extraction of a drug lead compound using molecularly imprinted polymers prepared by the target analogue approach. Chromatographia 56:237–241
92. Zhu QZ, Haupt K, Knopp D, Niessner R (2002) Molecularly imprinted polymer for metsulfuron-methyl and its binding characteristics for sulfonylurea herbicides. Anal Chim Acta 468:217–227
93. El-Toufaili FA, Visnjevski A, Bruggemann O (2004) Screening combinatorial libraries of molecularly imprinted polymer films casted on membranes in single-use membrane modules. J Chromatogr B 804:135–139
94. Davies MP, de Biasi V, Perrett D (2004) Approaches to the rational design of molecularly imprinted polymers. Anal Chim Acta 504:7–14
95. Navarro-Villoslada F, San Vicente B, Moreno-Bondi MC (2004) Application of multivariate analysis to the screening of molecularly imprinted polymers for bisphenol A. Anal Chim Acta 504:149–162

96. Koohpaei AR, Shahtaheri SJ, Ganjali MR, Forushani AR, Golbabaei F (2008) Application of multivariate analysis to the screening of molecularly imprinted polymers (MIPs) for ametryn. Talanta 75:978–986
97. Xu D, Zhu W, Jiang Y, Li XS, Li WN, Cui JC, Yin JX, Li GT (2012) Rational design of molecularly imprinted photonic films assisted by chemometrics. J Mater Chem 22:16572–16581
98. Muhammad T, Nur Z, Piletska EV, Yimita O, Piletsky SA (2012) Rational design of molecularly imprinted polymer: the choice of cross-linker. Analyst 137:2623–2628
99. Muhammad T, Liu C, Wang JD, Piletska EV, Guerreiro AR, Piletsky SA (2012) Rational design and synthesis of water-compatible molecularly imprinted polymers for selective solid phase extraction of amiodarone. Anal Chim Acta 709:98–104
100. Nantasenamat C, Naenna T, Ayudhya CIN, Prachayasittikul V (2005) Quantitative prediction of imprinting factor of molecularly imprinted polymers by artificial neural network. J Comput Aid Mol Des 19:509–524
101. Kempe H, Kempe M (2004) Novel method for the synthesis of molecularly imprinted polymer bead libraries. Macromol Rapid Commun 25:315–320
102. Rosengren AM, Karlsson JG, Andersson PO, Nicholls IA (2005) Chemometric models of templatesmolecularly imprinted polymer binding. Anal Chem 77:5700–5705
103. Rosengren AM, Golker K, Karlsson JG, Nicholls IA (2009) Dielectric constants are not enough: principal component analysis of the influence of solvent properties on molecularly imprinted polymer-ligand rebinding. Biosens Bioelectron 25:553–557
104. Golker K, Karlsson BCG, Rosengren AM, Nicholls IA (2014) A functional monomer is not enough: principal component analysis of the influence of template complexation in pre-polymerization mixtures on imprinted polymer recognition and morphology. Int J Mol Sci 15:20572–20584
105. Piletsky SA, Karim K, Piletska EV, Day CJ, Freebairn KW, Legge CH (2001) Recognition of ephedrine enantiomers by molecularly imprinted polymers designed using a computational approach. Analyst 126:1826–1830
106. Wei S, Jakusch M, Mizaikoff B (2007) Investigating the mechanisms of 17-estradiol imprinting by computational prediction and spectroscopic analysis. Anal Bioanal Chem 389:423–431
107. Henschel H, Kirsch N, Hedin-Dahlstrom J, Whitcombe MJ, Wikman S, Nicholls IA (2011) Effect of the cross-linker on the general performance and temperature dependent behaviour of a molecularly imprinted polymer catalyst of a Diels-Alder reaction. J Mol Catal B-Enzym 72:199–205
108. Henthorn DB, Peppas NA (2007) Molecular simulations of recognitive behavior of molecularly imprinted intelligent polymeric networks. Ind Eng Chem Res 46:6084–6091
109. Molinelli A, O'Mahony J, Nolan K, Smyth MR, Jakusch M, Mizaikoff B (2005) Analyzing the mechanisms of selectivity in biomimetic selfassemblies via IR and NMR spectroscopy of prepolymerization solutions and molecular dynamics simulations. Anal Chem 77:5196–5204
110. Ahmadi F, Rezaei H, Tahvilian R (2012) Computational-aided design of molecularly imprinted polymer for selective extraction of methadone from plasma and saliva and determination by gas chromatography. J Chromatogr A 1270:9–19
111. Ahmadi F, Ahmadi J, Rahimi-Nasrabadi M (2011) Computational approaches to design a molecular imprinted polymer for high selective extraction of 3,4-methylenedioxymethamphetamine from plasma. J Chromatogr A 1218:7739–7747
112. Ahmadi F, Yawari E, Nikbakht M (2014) Computational design of an enantioselective molecular imprinted polymer for the solid phase extraction of Swarfarin from plasma. J Chromatogr A 1338:9–16
113. Puzio K, Delepee R, Vidal R, Agrofoglio LA (2013) Combination of computational methods, adsorption isotherms and selectivity tests for the conception of a mixed non-covalent-semi-covalent molecularly imprinted polymer of vanillin. Anal Chim Acta 790:47–55
114. Martins N, Carreiro EP, Locati A, Ramalho JPP, Cabrita MJ, Burke AJ, Garcia R (2015) Design and development of molecularly imprinted polymers for the selective extraction of

deltamethrin in olive oil: an integrated computational-assisted approach. J Chromatogr A 1409:1–10
115. Gholivand MB, Khodadadian M, Ahmadi F (2010) Computer aided-molecular design and synthesis of a high selective molecularly imprinted polymer for solid-phase extraction of furosemide from human plasma. Anal Chim Acta 658:225–232
116. Nezhadali A, Shadmehri R (2014) Neuro-genetic multi-objective optimization and computer-aided design of pantoprazole molecularly imprinted polypyrrole sensor. Sensor Actuat B Chem 202:240–251
117. Nezhadali A, Pirayesh S, Shadmehri R (2013) Computer-assisted to sensor design and analysis of 2-aminobenzimidazole in biological model samples based on electropolymerized-molecularly imprinted polypyrrole modified pencil graphite electrode. Sensor Actuat B Chem 185:17–23
118. Prasad BB, Rai G (2012) Study on monomer suitability toward the template in molecularly imprinted polymer: an ab initio approach. Spectrochim. Acta A 88:82–89
119. Lulinski P, Sobiech M, Zoek T, Maciejewska D (2014) A separation of tyramine on a 2-(4-methoxyphenyl)ethylamine imprinted polymer: an answer from theoretical and experimental studies. Talanta 129:155–164
120. Gholivand MB, Khodadadian M (2011) Rationally designed molecularly imprinted polymers for selective extraction of methocarbamol from human plasma. Talanta 85:1680–1688

Chapter 3
Special Control by Molecularly Imprinted Materials-Zero-Order Sustained Release, Enantioselective MIPs, and Self-Regulated Drug Delivery Microdevices

Xue Zhang, Xiao Liu, Ze-Hui Wei, and Yan-Ping Huang

3.1 Molecularly Imprinted Materials-DDS for Zero-Order Sustained Release

Compared to the other conventional dosage forms, zero-zero release systems have the advantages that can keep drug concentrations in the body at optimal levels, so they have been developed and still in development. Based on previous studies, such as Fick's law of diffusion, it is an inherently nonlinear phenomenon that the drug release from matrix [1]. To overcome these shortcomings, the molecularly imprinted polymers with specific recognition function were prepared by mimicking the recognition behavior of natural receptors because the molecular imprinting has a specific recognition performance for template and can be used as a special material in the design of drug delivery system [2, 3]. In recent years, the potential application of MIPs in drug delivery system has received much attention. But generally speaking, the zero-level release system of MIPs is not available, because how MIPs morphology and affinity are controlled by aggregation parameters is not clear [4].

Tang et al. reported a molecular crowding strategy for the preparation of the facile fabrication of zero-order sustained release systems by molecular imprinting technique. Aminoglutethimide (AG) was selected as a model drug, and the crowding-assisted MIPs matrices were synthesized by free-radical precipitation in this report. The purpose of improving imprinting effect is achieved by adding polystyrene (PS) as a polymer co-solute in the pre-polymerization solution. The key factor of the polymerization variables is the porogen system of PS-THF, because the volume

X. Zhang · Z.-H. Wei · Y.-P. Huang (✉)
College of Pharmacy, Tianjin Medicine University, Tianjin, China

X. Liu
Tianjin Huanhu Hospital, Tianjin, China

Z. Liu et al. (eds.), *Molecularly Imprinted Polymers as Advanced Drug Delivery Systems*, https://doi.org/10.1007/978-981-16-0227-6_3

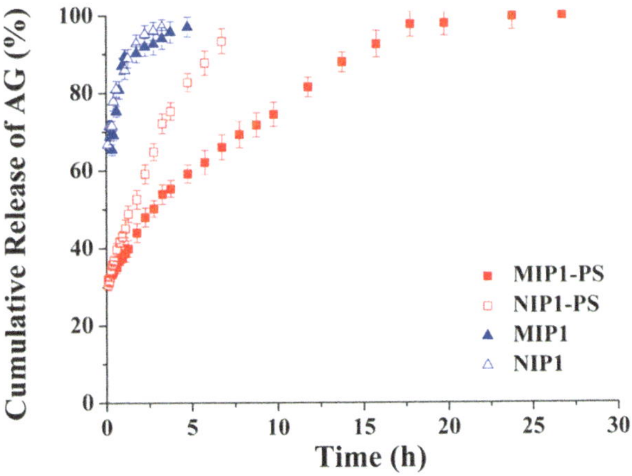

Fig. 3.1 Release curves of AG from PS-based MIP (MIP1-PS), PS-based NIP (NIP1-PS), PS-free MIP (MIP1) and PS-free NIP (NIP1). (Reproduced with permission of Ref. [5])

exclusion effect affects the molecular imprinting sites of the polymer network and MIPs [5]. Compared with the MIP under non-crowding condition, the corresponding imprinted particles with PS gave a markedly relatively long release time 20 h, whereas the PS-assisted NIP gave only 7.5 h. For both MIP and NIP without PS, AG was released at a very high rate (85% occurring in the previous hour). In this report, the MIPs under the crowding condition were successfully prepared for zero-order sustained release. To explore the MIP morphology, scanning electron micrography was used. The result showed that PS had not remarkable influence on the morphology (Fig. 3.1).

Ketotifen fumarate ($M_W = 425$), a low molecular weight drug, was released with zero order at the molecular level. It is also used as an imprinted hydrogel in therapeutic contact lenses. A dynamic in vitro release study of imprinted hydrogel contact lens was performed by Maryam's team, and the novel microfluidic device was designed to stimulate the volume flow rate, tear volume, and tear composition of the eye. The release spectrum of various components of gels was obtained by mapping the drug release classification (M_t/M_∞) and the time normalization to the gel thickness square (t/L2). As shown in the report, the binding concentration with ketotifen in gels are 4.9×10^{-2}, 1.7×10^{-2}, 7.4×10^{-3} and 5.1×10^{-3} mmol/g for poly(AA-co-AM-co-NVP-HEMA-PEG200DMA), poly(AA-co-AM-co-HEMA-PEG200DMA), poly(AM-co-HEMA-PEG200DMA) and poly(AA-co-HEMA-PEG200DMA), respectively. The release kinetic was zero order under physiological ocular volumetric flow rates. The release kinetics control of the swelling behavior of the polymer was extended by the imprinting process, which led to great changes in the diffusion coefficient [6].

3.2 Enantioselective Drug Delivery System and the Application of MIP

3.2.1 MIPs Used in Oral Drug Delivery Systems

Early studies of the use of MIPs for oral enantiomeric selective drug delivery systems were most tables. The significant difference of stereoisomers in pharmacological activity and pharmacokinetics also arouse people's attention with the development of chiral drugs in recent years. The release test showed that the imprinted enantiomers were released more slowly than other enantiomers. In general, one enantiomer is more efficient than the other. In this case, it is better to use only the more active enantiomers. Some enantiomers are biotoxic, so the selective release of enantiomers is necessary in oral preparations. However, the problems of preparing enantiomeric pure drugs are difficult to be overcome. For example, the method of preparing single enantiomer has many problems, such as asymmetric synthesis and chromatographic chiral separation, as well as high cost, difficulty and time [7].

Suedee et al. designed and prepared MIPs with *R*-propranolol as template. In this report, MIPs were prepared by using *R*-propranolol, *S*-buprofen, and *S*-ketoprofen as template, and the selective release effect of the MIPs particles on racemate of three chiral drugs was investigated under different pH conditions. The results showed that the release rate of all drugs used as template molecules was slower than that of their corresponding chiral isomers. As can be seen from Fig. 3.2, the drug polymer ratio of the *S*-ibuprofen MIP particles increased with decreasing the release percentage of ibuprofen enantiomers. The release percentage of 1:5 that was lower than at 1:1 ratio in the last dissolution stage after 30 h. In addition, approximately 15% and 10% of *R*-ibuprofen and *S*-ibuprofen cumulative releases were detected at the end of the dissolution run, respectively. The sustained release of enantiomers was at the ratio of 1: 25. The results showed that the MIP particles had a stronger inhibitory effect on ibuprofen enantiomer release than non-MIP particles. This rescult indicates that ibuprofen enantiomers are preferentially adsorbed on the MIP [8].

The aims of Suedee's research are to developing enantiome-controlled oral drug delivery systems to selectively release the required (*S*) enantiomers, including racemates in dose formulations that respond to pH stimuli. The identification system was taken by molecular imprinting of nanoparticles microspheres, and a multifunctional chiral cinchona anchor polymer was prepared by means of suspension polymerization. To make the polymer exhibit high stereoselectivity, ethylene glycol dimethacrylate and (*S*)-omeprazole were used as cross-linking agent and imprinted template, respectively. The results show that the microsphere delivery system with (*S*)-omeprazole imprinted cinchona polymer nanoparticles can maximize the efficacy and minimize the dose frequency. As shown in Fig. 3.3, nonspecific adsorption of omeprazole (MIP/NIP = 1) has a very high enantiomer buffer for MQN- and MQD-based MIPs in aqueous solution. This may be due to the fact that ester bonds in the polymer EDMA polymer network favor nonspecific reciprocities with the hydrophobic part of the drug. Those results show that pH 7.4

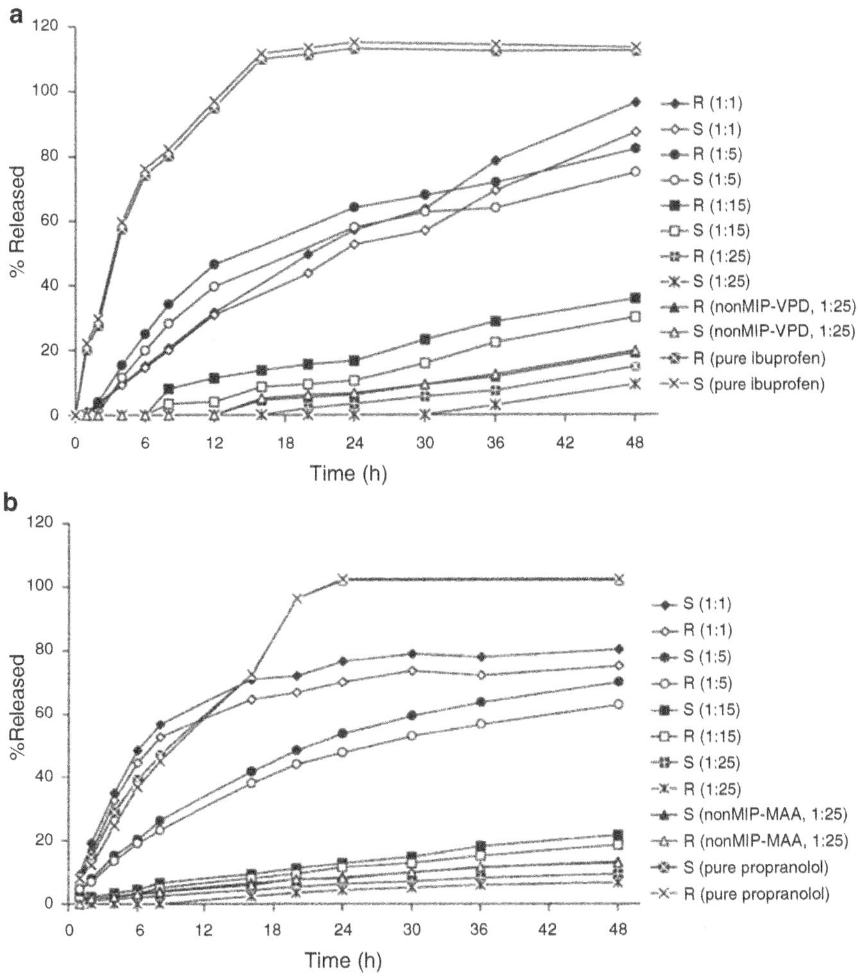

Fig. 3.2 Influence of drug polymer ratio on enantiomer release rate within 48 h from (**a**) *S*-ibuprofen MIP particle and (**b**) *R*-propranolol MIP particle at pH 7.4 and 37 °C. (Reproduced with permission of Ref. [8])

buffer has a good pH value and selectional binding of enantiomers to the polymer. These synthesized matrices can be used to develop systems that selectively control enantiomer release in response to the formation of racemic omeprazole sustained release matrices with varying pH values [9].

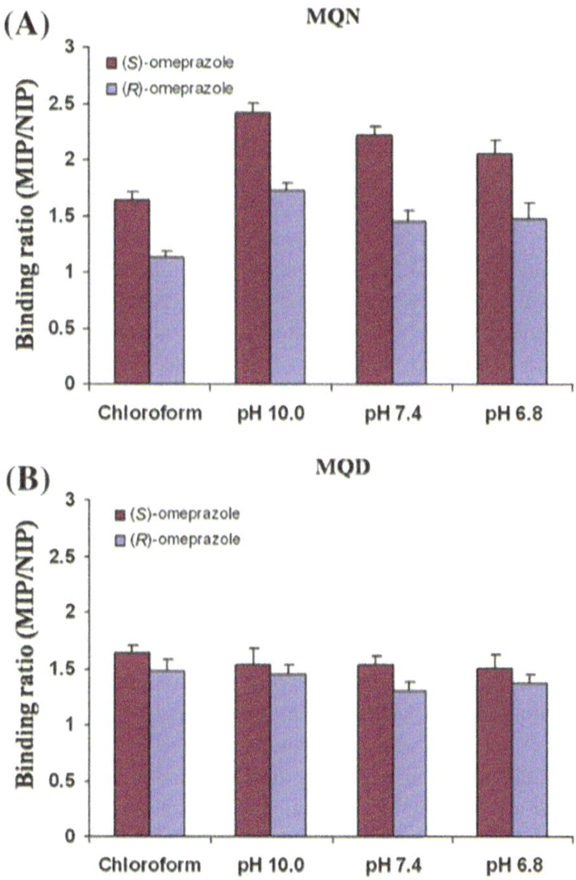

Fig. 3.3 At room temperature, omeprazole (*R*) and (*S*) enantiomers were compared with MIP-containing enantiomers when racemic omeprazole was incubated in chloroform or water buffer solutions (mean ± SD, $n = 3$). The nonspecific adsorption ratio of MIP/NIP was 1. The binding ratio of (*R*)- and (*S*)-enantiomers in a medium of pH 1.2 was not detectable. (Reproduced with permission of Ref. [9])

3.2.2 MIPs Used in Transdermal Drug Delivery Systems

Propranolol is not a selective epinephrine receptor antagonist compared to others. The effectiveness of oral administration is influenced by extensive first-generation metabolism and low absorption. Percutaneous administration of propranolol has the potential to improve bioavailability by avoiding the first pass effect of hepar. The pharmacological activity of *S*-enantiomer propranolol is 100–130 times that of *R*-enantiomer, as it has only one chiral center according to the chemical formula. The stereoselectivity of propranolol ester prodrug and propranolol was not obtained in vitro, so propranolol was marketed as racemate [10]. Therefore, the treatment results of S-enantiomers selectively applied to the skin surface were better than that of racemes mixture.

In this study, MIP was combined with bacteria-derived membranes to form a composite membrane, which is expected to be used as a model transdermal drug delivery system for the *S*-propranolol [11]. To ensure the polymer has a uniform distribution within the cellulose membrane pores, optimizing MIP affinity was studied. By minimizing nonselective diffusion through perforation, it is possible to achieve selective absorption of propranolol and subsequent release through the skin. On this basis, the selective molecularly imprinted photonic membranes of *S*-propranolol were developed and their release experiments were studied in vitro. Preliminary studies have shown that the initial concentration of 40 μg ml^{-1} of chlorophenol GML-1 was able to establish a steady state flux within 2–6 h (Fig. 3.4). The transfer rate of *R*-and *S*-protoluol enantiomers on the unmodified cellulose membrane was faster than that after modification, indicating that the modified cellulose membrane had the function of pore filling. Compared with *R*-enantiomer, *S*-enantiomer had a faster transmission speed, which investigated the intrinsic enantioselectivity of the revised cellulose membrane. Because the *S*-MIP membrane limited the penetration of *R*-enantiomers, they were not monitored in the receptor compartment for at least 6 h. However, the *R*-enantiomer still transferred slower than that of the *S*-enantiomer through *R*-MIP membrane. The *R*-isomers that passed through the *R*-MIP membrane were faster than through the NIP or *S*-MIP membranes, and this was due to the selective application of *R*-enantiomer to some membrane pores. In a nutshell, these results indicate that enantiomers can across the composite membrane in many ways.

A phase change technique using polypyruvate-triol as plasticizer was used. In order to form membranes with high permeability, initial design objectives included requiring template molecules (*S*-enantiomers of propranolol) to be highly close to binding sites on MIP particles. As shown in scanning electron micrographs (Fig. 3.5), (a), (b), and (c) were particles, beads and MIP-NOM, respectively [12]. The polymer granules were significantly different from microbeads in morphology, which was obviously dependent on the polymerization method. The monolithic particles that were taken by bulk polymerization, grinding, and sieving had a range of characteristics, including irregular and rough morphologies with diameters ranging from 15 to 35 μm. Due to the small size of nanoparticles, nano-imprinted polymers would be generally expected to supply large specific surface area with a high selectivity. However, it was difficult to prepare polymer nanoparticles by phase transition and subsequent casting into the composite membrane. On the other hand, these particles tend to present in the forming aggregates, and then reduce the effective specific surface area of the particles. The sustainable competitive advantages of the MIP-NOM were not only providing highly accessible imprinted sites, but also solving the processing problems associated with the generation and incorporation of nanoparticles.

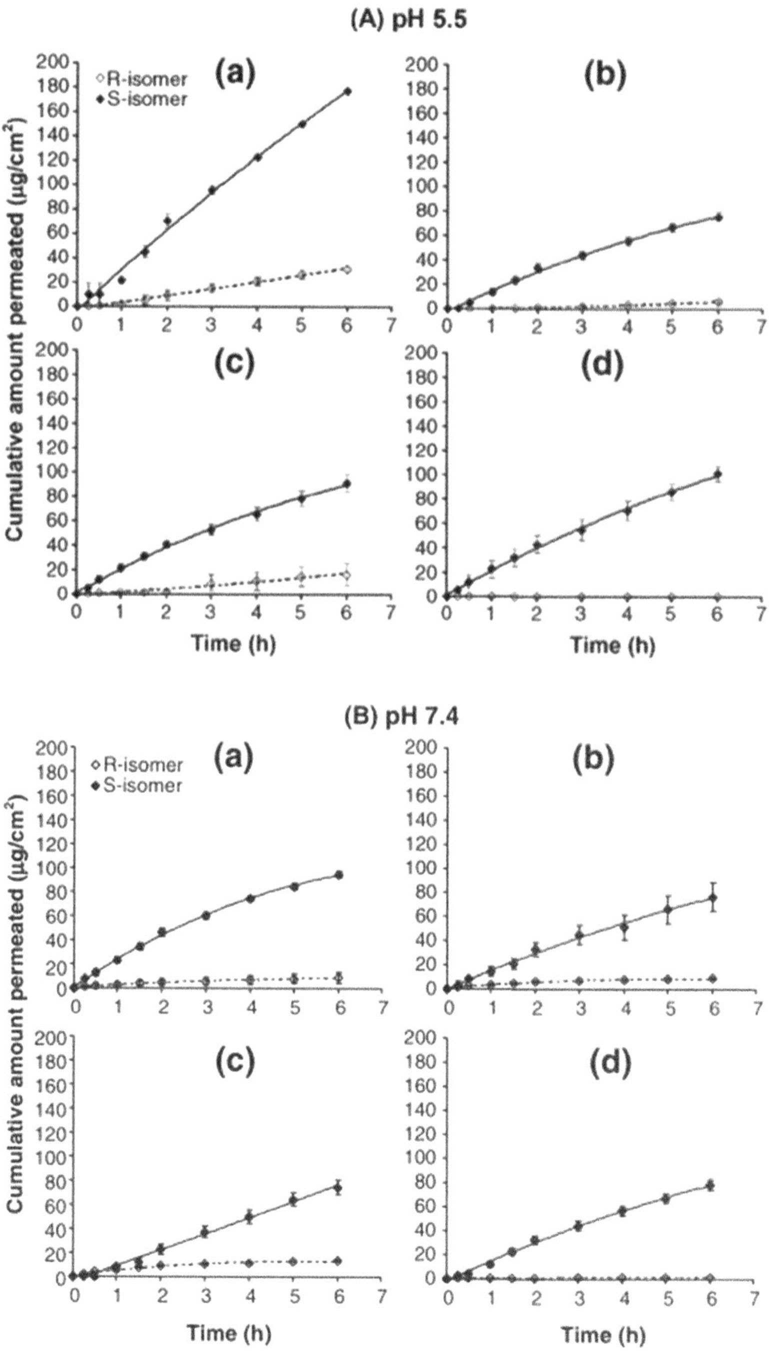

Fig. 3.4 The transport changes of *R*- and *S*-propranolol enantiomers (HCl) over time (mean ± SE, $n = 3$) from (**A**) pH 5.5 citrate buffer and (**B**) pH 7.4 phosphate buffer across (**a**) cellulose membrane, (**b**) NIP membrane, (**c**) *R*-MIP membrane and (**d**) *S*-MIP membrane. (Reproduced with permission of Ref. [11])

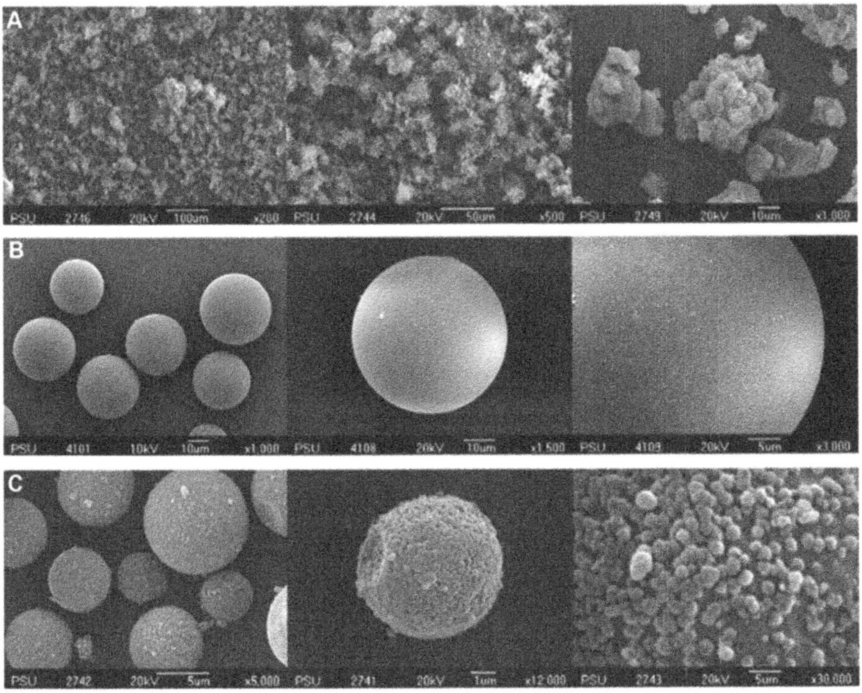

Fig. 3.5 SEM images of the imprinted polymer particles: granules (**a**), microspheres (**b**), nanoparticle-on-microspheres (**c**). (Reproduced with permission of Ref. [12])

3.3 Self-Regulated Drug Delivery Microdevices Based on MIP

3.3.1 Environment-Stimuli Responsive MIP

One of the most widely studied polymer materials was cross-linked N-substituted polyacrylamide, which was used in biological molecular imprinting of proteins, DNA, etc. [13]. The field of controlled delivery of drugs has received extensive attention, because these polymers could undergo phase transition of temperature control volume in aqueous solution. A promising strategy was provided by combining the capacities of thermosensitive polymers with molecular imprinting technology to ensure that the system responds more quickly to outer temperature changes. Copolymerization of acrylamide with cross-linking agent and additional monomer was used to synthesize imprinted polymers, and these polymers exhibited reversible phase transitions in the presence of templates at room temperature [14]. Silva et al. reported a novel pH-sensitive MIP with the recognition ability for ibuprofen. The potential drug delivery polymer that consisted of 2-(dimethylamine) ethyl methacrylate (DMAEMA) and ethylene glycol dimethacrylate (EGDMA) was successfully synthesized at 65 °C and 21 MPa with $scCO_2$ as solvent and poregenic agent. $ScCO_2$

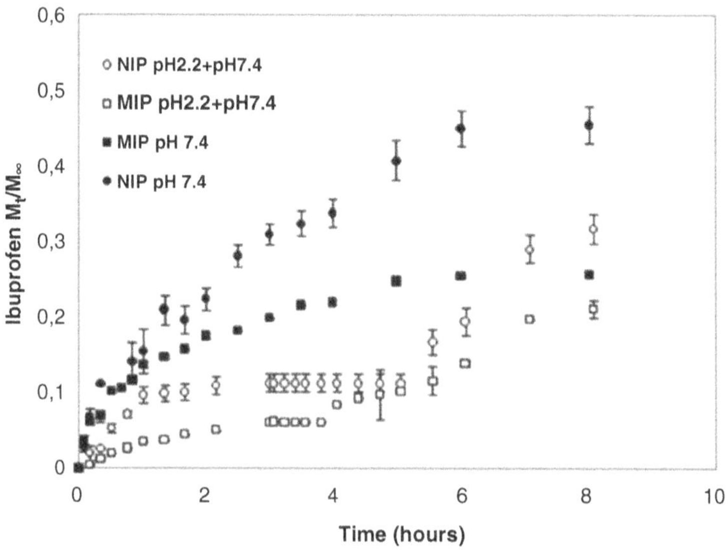

Fig. 3.6 Eight hours of ibuprofen in vitro release spectra were obtained from MIP and NIP at 37 °C and at different pH. Each result is the average of at least three separate experiments. (Reproduced with permission of Ref. [15])

impregnation method was used to prepare the drug, and the template was desorbed in supercritical environment. In vitro drug delivery experiments have demonstrated that it is possible to obtain polymer substrates with high drug affinity and sustained drug release, even at low cross-linking rates. Figure 3.6 shows an experimental profile of time-varying ibuprofen release from MIP and NIP samples. The substrate was exposed to pH 7.4 for 8 h or pH 2.2 for 3 h and then pH 7.4 for 5 h. Because ibuprofen has a low solubility at an acidic pH, its release was hindered by its relative release at a neutral pH [15].

Va-imprinted nanospheres sensitive to pH were prepared by UV-induced precipitation polymerization. Vancomycin (VA) was used as a template in this study. As is shown in Fig. 3.7, 2-hydroxyethyl methacrylate (HEMA) and DEAEMA are used as bifunctional monomers, and EGDMA is used as a cross-linking agent. The prepared VA@MIPs can be significantly influenced to pH. For example, compared with neutral conditions, a faster release rate and a higher release concentration could be achieved under acidic conditions. This is necessary to inhibit the bacteria infections. On the other hand, these MIPs not only have a slower release than NIPs, but also release longer than 18 days [16].

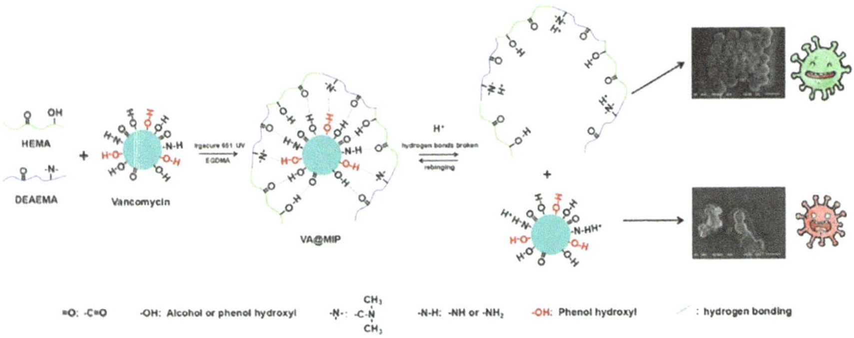

Fig. 3.7 Imprinting mechanism for the drug delivery system of VA@MIPs. (Reproduced with permission of Ref. [16])

Table 3.1 Extent of uptake of T and Hy by MIP and the control polymer

Compound	Amount absorbed by 100 mg MIP (μg)	Amount absorbed by 100 mg of control polymer (μg)
Testosterone	175 ± 4	36 ± 2
Hydrocortisone	296 ± 5	46 ± 3

Reproduced with permission of Ref. [17]

3.3.2 Chemical-Stimuli Responsive MIP

If the structure of the drug molecule is similar to the template, the resulting MIPs can also selectively adsorb the drug. Drug release from MIPs is affected by the presence of template molecules in the dissolution medium, because there is competitive binding between template molecules and MIPs drugs. Template molecules will "replace" the drug from the drug-loaded MIPs and accelerate its release. For example, testosterone (T) and hydrocortisone (Hy) have very similar molecular structures, and hydrocortisone is used as a template molecule to prepare MIPs. The molecularly imprinted polymers with 2-hydroxy-ethyl-methacrylate were prepared to absorb significant amounts of testosterone. This makes it possible to modulate the release of specific molecules in the drug delivery system. Table 3.1 summarizes the degree of absorption for Hy and T by MIP and control. Compared with non-imprinted polymers, MIPs absorb a large amount of T. The molecular size of T is close to Hy, which may contribute to the upregulation of MIP in this molecule level [17]. Thus, such a molecularly imprinted drug release system can be applied to steroids and peptides.

Although the use of MIPs in this field is still in the fledging period, the application of MIPs in the design of new DDS and related devices, such as diagnostic sensors, is now arousing more and more interest. The MIP-based DDS provide three main methods for controlling the initiation time of administration and/or the rate of drug release, activation regulation, rate program or feedback regulation of administration [18].

In different intellectual stimuli-response platforms, cross-linked hydrogel still is a good choice. Hydrogels can be prepared from cross-linked stimuli-sensitive polymers by polymerization, or from monomers and cross-linking agent that are cross-linked in a subsequent step. The detailed hydrogel preparation for protocol can be found in other reports [19].

How do drugs modify THE sensitivity of DDS to treatment-related stimuli (stimulus thresholds) and responsiveness (drives) that has been discussed in this study. The drug release patterns were changed by elucidating the different drug binding patterns of swelling and collapse states [26]. This information suggests that drug loading and release may trigger phase transitions in non-drug-responsive hydrogels (i.e., a priori non-analytically responsive network). Better understanding the effects of drugs on reactivity is a necessary step in the clinical application of smart hydrogels, and it is also possible to unveil new uses for stimulating reactivity of DDS [20].

3.3.3 Application of MIPs in Target Drug Delivery System

MIPs have a wide range of applications in different scientific fields, but the remarkable thing is that their most attractive applications are therapeutics and medical therapies [21]. However, the drugs currently on the market are difficult to balance with targeting, diagnosis, and delivery functions. These characteristics are necessary for reducing unexpected drug leakage in vivo and accurated diagnosis of the drug delivery system. For example, when MIP binds to a target on the cell surface, the drugs that are covalently or non-covalently bound to the MIP for targeted drug delivery are released [22].

According to previous reports, MIP-polyethylene glycol-folic acid (MIP-PEG-FA) nanoparticles have been successfully prepared, and then used as a controlled release carrier of paclitaxel (PTX) targeted to cancer cells. The MIP nanoparticles were synthesized by microemulsion polymerization, and then PEG-FA was connected to the surface of nanoparticles. Cytotoxicity of PTX, MIP, and MIP-PEG-FA against MDA-MB-231(folate positive) and A549 (folate negative) cancer cells was tested in this study. The MIP nanoparticles containing PTX have better uptake ability to FA receptor-positive cell lines than non-targeted MIP nanoparticles. Imprinted polymer nanoparticles with high drug loading showed stronger cytotoxicity than pure PTX nanoparticles [23].

Although single-template molecularly imprinted polymers (SMPS) have been widely used in drug delivery, they often fail to deliver drugs and target them precisely at the same time. Two-template molecularly imprinted polymer nanoparticles were prepared for the targeted diagnosis and administration of pancreatic cancer BXPC-3 cells (FH-MIPNPS) by Jia's team [24]. In FH-MIPNPs, the 71–80 peptide of human fibroblast growth factor induction factor 14 modified by

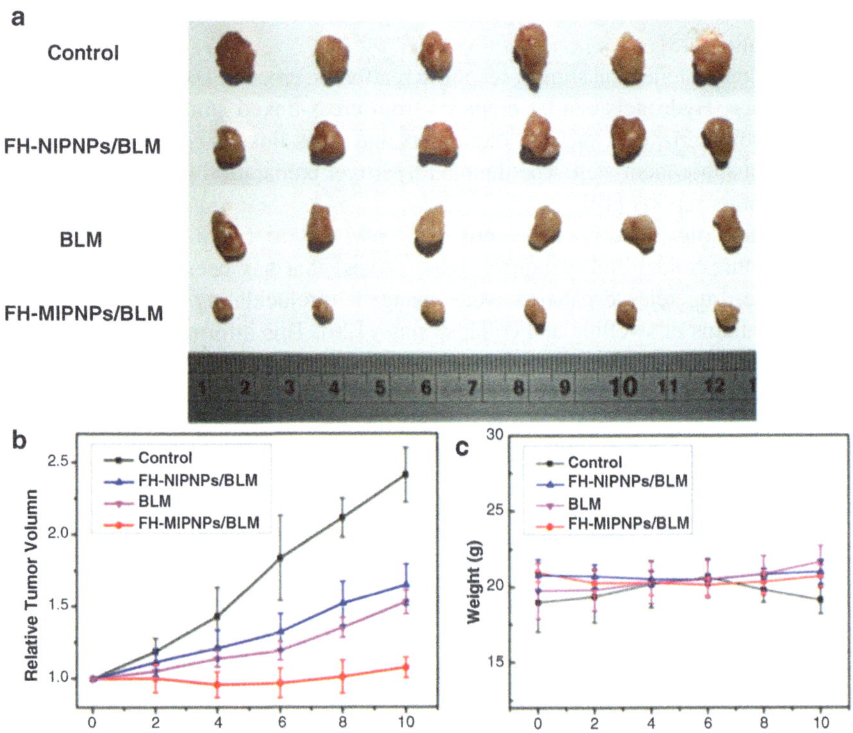

Fig. 3.8 (**a**) Digital photos showing four groups of tumor morphology (**b**) In the control group (normal saline), BXPC-3 tumor-bearing nude mice were treated with FH-MIPNPS /BLM, FH-NIPNPS /BLM and BLM. (**c**) Weight of mice in each group (Reproduced with permission of Ref. [24])

glucose (Glu-FH) and bleomycin (BLM) were used as template. At the same time, FH-MIPNPS was loaded with BLM and bound to BxPC3 cells that overexpressed human fibroblast growth factor induction factor 14 (FN14). As shown in Fig. 3.8, FH-MIPNP-loaded BLM (FH-MIpNPS/BLM) could inhibit the growth of xenograft tumor BXPC-3 (the tumor volume increased to 1.05 times) from in vivo antitumor tests. Targeted delivery of drugs can be achieved by the specific recognition function of MIPs. The results of cytotoxicity test and tissue section staining showed that the toxicity was low.

3.3.4 Application of MIPs in Drug Trap System

In order to prevent some substances that are not absorbed by the body, and the specificity of MIPs for certain substances are used to trap these substances for the purpose of curing the disease. The target substance is often something related to disease or toxic substances, the most studied in recent research are glucose and cholesterol capture system.

Purification of these MIP hydrogels may lead to a product to help treat type 2 diabetes mellitus. MIPs could be used as a drug, and bind glucose in the stomach and small intestine and pass through the undigested parts of the body.

The drug can be taken simply with high-sugar foods, thus reducing the sharp rise in blood sugar caused by consuming large amounts of simple sugars [25]. The cholesterol imprinted poly (2-hydroxyethyl methacrylate-methylacrylamide tryptophan) (PHEM-MTRP) particles were prepared to form embedded composite membranes in the recent research. The cholesterol selectivity of the imprinted membrane was 1.96 and 2.13 times higher than that of the competitors including stigmasterol and estradiol, respectively. According to Mehmet Odabas's report, when the cholesterol concentration increased by 2 mg/mL, the cholesterol adsorption capacity increased to 23.43 mg/g. The MIP-coated composite membranes have little effect for the adsorption of cholesterol, after using the same adsorbent for ten consecutive times of adsorption. As a result, it's a reliable, new, easy-to-use, and an inexpensive treatment. There is an urgent need to remove cholesterol from the hyped cholesterolemic blood based on the sensitive and selective systems [26].

In recent years, a new-style molecular imprinting technique was presented by Gao's team, and uric acid-imprinting material (MIP-PAO/CPVA) with high performance was successfully obtained using CPVA microspheres as substrate. MIP-PAO /CPVA not only has high affinity and specificity for template binding, but also has biocompatibility. It is a promising absorbent of solid uric acid for blood purification. As shown in Fig. 3.9, the whole preparation process of MIP-PAO/CPVA and the potential binding model of graft PAO macromolecule and uric acid molecule were presented. The binding amount of the polymer to uric acid is much higher than that to guanine from Fig. 3.10. The experimental results show that a better template binding effect can be obtained by strong bond interaction. The novel surface imprinting method proposed can be used to prepare various high-performance adsorbents for blood purification, and macromolecule microspheres with biocompatibility or blood compatibility are used as matrix. Therefore, the results of this study provide valuable reference for the development of blood purification materials [27].

3.4 Conclusion

The application progress of MIPs in drug zero-order sustained release and the application of enantioselective intelligent drug release system has been summarized in this chapter. MIPs can regulate drug release by increasing residence time in the polymeric matrix through covalent or non-covalent interactions at specific binding sites. Therefore, MIPs are promising materials in the building of drug release

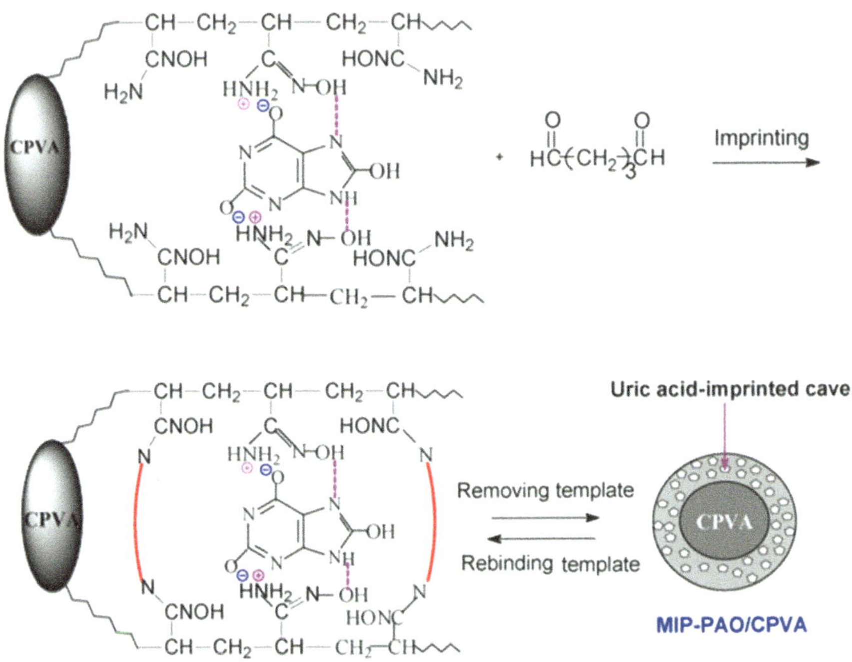

Fig. 3.9 Schematic expression of preparation process of MIP-PAO/CPVA. (Reproduced with permission of Ref. [27])

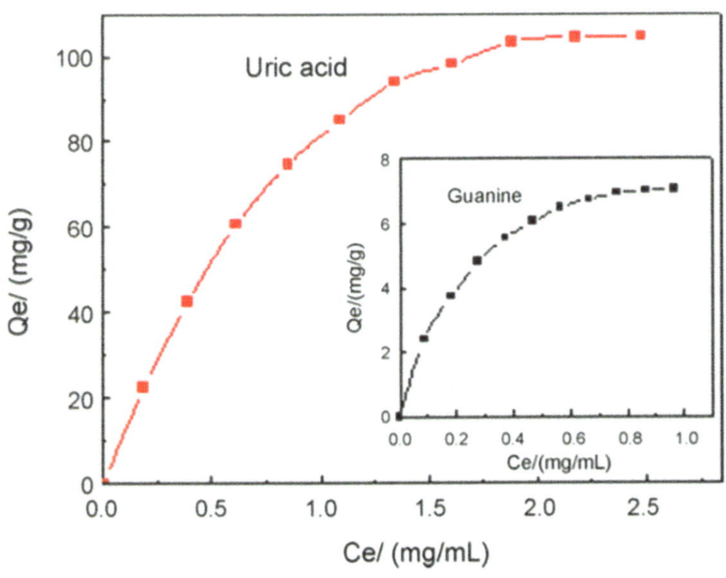

Fig. 3.10 Binding isotherms of MIP-PAO/CPVA for uric acid and guanine. Temperature: 30 °C; pH 7. (Reproduced with permission of Ref. [24])

appliance, because they can provide a better release spectrum and longer release time and release the drug in a feedback-modulated way, which is very important in modern drug therapy. With the development of molecular imprinting technology, it is foreseeable that there will be more MIPs drug delivery systems in the near future. The intelligent drug delivery system with biometric function designed by MIPs will play an increasingly important role in the treatment of diseases.

References

1. Chidambaram N, Porter W, Flood K, Qiu Y (1998) Formulation and characterization of new layered diffusional matrices for zero-order sustained release. J Control Release 52:149–158
2. Varelas CG, Dixon DG, Steiner CA (1995) Zero-order release from biphasic polymer hydrogels. J Control Release 34:185–192
3. Canfarotta F, Lezina L, Guerreiro A, Czulak J, Petukhov A, Daks A, Smolinska-Kempisty K, Poma A, Piletsky S, Barlev NA (2018) Specific drug delivery to cancer cells with double-imprinted nanoparticles against epidermal growth factor receptor. Nano Lett 18:4641–4646
4. Rostamizadeha K, Vahedpourb M, Bozorgib S (2012) Synthesis, characterization and evaluation of computationally designed nanoparticles of molecular imprinted polymers as drug delivery systems. Int J Pharm 424:67–75
5. Tang L, Zhao CY, Wang XH (2015) Macromolecular crowding of molecular imprinting: a facile pathway to produce drug delivery devices for zero-order sustained release. Int J Pharm 2:496–503
6. Ali M, Horikawa S, Venkatesh S, Saha J, Hong JW, Byrne ME (2007) Zero-order therapeutic release from imprinted hydrogel contact lenses within in vitro physiological ocular tear flow. J Control Release 124:154–162
7. Solin MA, Lugará S, Calvo B, Hernández RM, Gascón AR, Pedraz JL (1998) Release of salbutamol sulfate enantiomers from hydroxypropylmethylcellulose matrices. Int J Pharm 161:37–43
8. Suedee R, Srichana T, Rattananont T (2002) Enantioselective release of controlled delivery granules based on molecularly imprinted polymers. Drug Deliv 9:19–30
9. Suedee R, Jantarat C, Lindner W, Viernstein H, Songkro S, Srichana T (2010) Development of a pH-responsive drug delivery system for enantioselective-controlled delivery of racemic drugs. J Control Release 142:122–131
10. Riddell JG, Harron DW, Shanks RG (1987) Clinical pharmacokinetics of beta-adrenoceptor antagonists. An update. Clin Pharmacokinet 12:305–320
11. Danion A, Brochu H, Martin Y, Vermette P (2007) Fabrication and characterization of contact lenses bearing surface-immobilized layers of intact liposomes. J Biomed Mater Res A 82:41–51
12. Karlgard CC, Wong NS, Jones LW, Moresoli C (2003) In vitro uptake and release studies of ocular pharmaceutical agents by silicon containing and p-HEMA hydrogel contact lens materials. Int J Pharm 257:141–151
13. Slinchenko A, Rachkov H, Miyachi M, Ogiso M, Minoura N (2004) Imprinted polymer layer for recognizing double-stranded DNA. Biosens Bioelectron 20:1091–1010
14. Suedee R, Seechamnanturakit V, Canyuk B, Ovatlarnporn C, Martin GP (2006) Temperature sensitive dopamine-imprinted (N,N-methylene-bis-acrylamide cross-linked) polymer and its potential application to the selective extraction of adrenergic drugs from urine. J Chromatogr A 1114:239–249
15. Da Silva MS, Viveiros R, Morgado PI, Aguiar-Ricardo A, Correia IJ, Casimiro T (2011) Development of 2-(dimethylamino)ethyl methacrylate-based molecular recognition devices for controlled drug delivery using supercritical fluid technology. Int J Pharm 416:61–68

16. Mao C, Xie X, Liu X, Cui Z, Yang X, Yeung KWK, Pan H, Chu PK, Wu S (2017) The controlled drug release by pH-sensitive molecularly imprinted nanospheres for enhanced antibacterial activity. Mater Sci Eng C Mater Biol Appl 77:84–91
17. Sreenivasan K (1999) On the application of molecularly imprinted poly (HEMA) as a template responsive release system. J Appl Polym Sci 71:1819–1821
18. Angel C (2004) Molecularly imprinted polymers for drug delivery. J Chromatogr B 804:231–245
19. Alvarez-Rivera F, Concheiro A, Alvarez-Lorenzo C (2018) Epalrestat-loaded silicone hydrogels as contact lenses to address diabetic-eye complications. Eur J Pharm Biopharm 122:126–136
20. Alvarez-Lorenzo C, Grinberg VY, Burova TV, Concheiro A (2020) Stimuli-sensitive cross-linked hydrogels as drug delivery systems: impact of the drug on the responsiveness. Int J Pharm 579:119157
21. Allender CJ, Richardson C, Woodhouse B, Heard CM, Brain KR (2000) Pharmaceutical applications for molecularly imprinted polymers. Int J Pharm 195:39–43
22. Esfandyari-Manesh M, Javanbakht M, Atyabi F, Dinarvand R (2012) Synthesis and evaluation of uniformly sized carbamazepine-imprinted microspheres and nanospheres prepared with different mole ratios of methacrylic acid to methyl methacrylate for analytical and biomedical applications. J Appl Polym Sci 125:1804–1813
23. Esfandyari-Manesh M, Darvishi B, Ishkuh FA, Shahmoradi E, Mohammadi A, Javanbakht M, Dinarvand R, Atyabi F (2016) Paclitaxel molecularly imprinted polymer-PEG-folate nanoparticles for targeting anticancer delivery: characterization and cellular cytotoxicity. Mater Sci Eng C Mater Biol Appl 62:626–633
24. Jia C, Zhang M, Zhang Y, Ma ZB, Xiao NN, He XW, Li WY, Zhang YK (2019) Preparation of dual-template epitope imprinted polymers for targeted fluorescence imaging and targeted drug delivery to pancreatic cancer BxPC-3 cells. ACS Appl Mater Interfaces 11:32431–32440
25. Wizeman WJ, Kofinas P (2019) Molecularly imprinted polymer hydrogels displaying isomerically resolved glucose binding. Biomaterials 22:1485–1491
26. Odabaşı M, Uzun L, Baydemir G, Aksoy NH, Acet Ö, Erdönmez D (2018) Cholesterol imprinted composite membranes for selective cholesterol recognition from intestinal mimicking solution. Colloids Surf B Biointerfaces 163:266–274
27. Gao B, Liu S, Li Y (2010) Preparation and recognition performance of uric acid-imprinted material prepared with novel surface imprinting technique. J Chromatogr A 1217:2226–2236

Chapter 4
Water Compatible Molecularly Imprinted Polymers

Qiliang Deng

Molecular imprinting is a technique that synthesizes materials with recognition sites for the specific molecule. Molecularly imprinted polymers (MIPs) are usually also called as "plastic antibody." At present, MIPs as specific recognition elements have been extensively employed in various scientific and engineering fields such as separation, enrichment, sensor, catalytic, cancer therapy, and drug delivery. MIPs are usually prepared by co-polymerizing of a functional monomer and a cross-linker in the presence of target molecules (template molecule). When template molecules are removed from the materials, the recognition cavity structurally matched with the distribution of functional groups of the template molecules are left within MIPs. MIPs for the specific recognition low-molar mass molecule, peptides, protein, cell, virus, and bacteria have been reported. The interaction between the recognition cavity and the target molecule within MIPs delays the transfer rate of the template leaving from recognition sites, which can extend the retention time of the target molecule in MIPs. MIPs have been widely reported in the field of drug delivery, various drugs such as gatifloxacin, dipyridamole, prednisolone acetate, timolol maleate, citalopram, tramadol, nitroglycerine, risperidone, glycyrrhizic acid, ocular peptide drugs [1], targeting anticancer agents [2, 3], and oral insulin [4] have been involved in the field of drug release. Several papers have reviewed the advance of MIPs in drug delivery [5–7].

Traditional MIPs are usually produced in non-polar organic solvents because the interaction between functional monomer and the template mainly relies on non-covalent interaction such as electrostatic, ionic bonds, hydrogen bonds, and hydrophobic effect. The aqueous environment is usually un-beneficial for the realization of the affinity of MIPs, however, which is the necessary solvent system for

Q. Deng (✉)
College of Chemical Engineering and Materials Science, Tianjin University of Science and Technology, Tianjin, People's Republic of China
e-mail: yhdql@tust.edu.cn

Z. Liu et al. (eds.), *Molecularly Imprinted Polymers as Advanced Drug Delivery Systems*, https://doi.org/10.1007/978-981-16-0227-6_4

the drug delivery. Thus, water compatible MIPs have received much attention in recent years. In this chapter, the advance of water compatible MIPs are summarized.

4.1 pH-Sensitive Hydrogel-Based Molecularly Imprinted Polymers (HydroMIPs)

Hydrogels are constituted of a cross-linker and a polymer backbone, which easily bind water, and are insoluble in aqueous medium or biological fluids. Hydrogels can change physical and chemical property as the external factors change such as light, pH, ultrasonic, and temperature, and can be used as carrier for the controllable drug release by modulating a specific external stimulus. Hydrogels based MIPs combine the excellent biocompatibility of hydrogels and the highly specific recognition property of MIPs. MIPs hydrogel with pH-responsive ability have been developed to improve the loading and release ability [8].

MIPs with pH-responsive property are usually prepared with acidic or ionic functional monomers such as methacrylic acid, acrylic acid and 2-(diethylamino) ethyl methacrylate. Gong's group [9] developed a MIPs delivery system based on 2-(diethylamino) ethyl methacrylate. The ionic interaction between functional monomer and dexamethasone-21 phosphate disodium endows such MIPs system with the property of the pH-responsive release. Dexamethasone-21 phosphate disodium has two pK_a values (pK_{a1} 1.89 and pK_{a2} 6.4). Mono-anion and di-anion are coexisted as the equal molar ratio at pH 6.4. 2-(Diethylamino) ethyl methacrylate (pK_a 10.4) is cationic at pH 10.4. Dexamethasone-21 phosphate disodium interacts with the imprinting cavity through two pairs of ionic interactions in neutral or basic medium. As the pH value changed from neutral to acidic, more and more target molecules are protonated, which lead to a weaker ionic interaction, and subsequently a faster release rate. The MIP system showed a controlled Dexamethasone-21 phosphate disodium release over one and a half month. A pH-sensitive MIPs nanospheres for the release of vancomycin has been produced using 2-hydroxyethyl methacrylate (HEMA) and 2-(diethylamino) ethyl methacrylate (DEAEMA) as co-functional monomers and ethylene glycol dimethacrylate (EGDMA) as a cross-linker via a UV-initiated precipitation polymerization method. The obtained MIPs nanospheres displayed prominent property such as more template binding, slowly release and sensitive responsive to pH [10].

Recently, hydrogel materials based on dopamine were widely exploited due to the facile preparation process, outstanding mechanical property, and excellent biological compatibility. The active amino group of dopamine provides many opportunities to interact with other molecules containing carboxyl, carbonyl, hydroxyl, and aldehyde groups. Ciprofloxacin possesses carbonyl group, carboxyl, and amino groups, which easily interact with amino group of dopamine. Wang et al. prepared a pH-responsive hydrogel for the release ciprofloxacin based on the combination of cellulose and

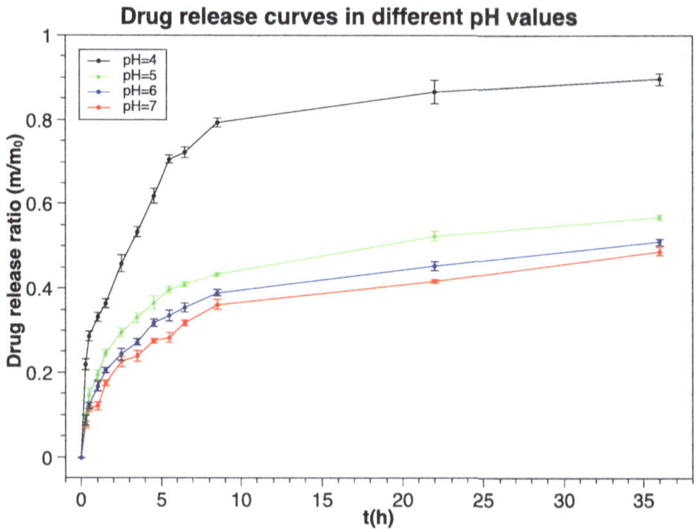

Fig. 4.1 Drug release curves of hydrogel under different pH medium. (Reproduced with permission of Ref. [11])

dopamine. The resulted material demonstrated the loading efficiency of 80.56%, pH-responsive ability, and long-term bacteriostatic property [11] (Fig. 4.1).

In a recent report, a methacrylic acid functionalized dopamine was also utilized to develop the drug release system. PH-responsive MIPs hydrogel for the release of dexamethasone sodium phosphate was developed using dopamine acrylamide and acrylic acid as co-functional monomers and N,N-methylene-bisacrylamide as a cross-linker via precipitation polymerization. The maximum binding amount of the resulted MIPs to template molecule is 28 mg g^{-1}. The system sustains the low release rate under acidic and alkaline conditions. The release rate is accelerated under weak acid conditions due to the destruction of hydrogen bonds interaction. The MIPs loaded with dexamethasone sodium phosphate can lastly release 83% of the binding mass within 7 days [12] (Fig. 4.2).

Sadegh and coworkers [13] prepared the three kinds MIPs via bulk polymerization, where methacrylic acid, 4-vinyl pyridine and methacrylamide were used as monomer, respectively, and ethylene glycol dimethacrylate as cross-linker. Furthermore, authors compared the binding and release properties of the three kinds MIPs for drug delivery of diclofenac in different pH values. Although the weaker affinity was observed in water than in organic medium, all MIPs worked for pH-responsive diclofenac delivery. Among these MIPs, the materials prepared by methacrylic acid showed superior properties, where the system showed minimum release (14%) in gastric acid and maximum release (90%) in basic condition. Recently, thermodynamic computational calculations based on Hansen method were also used to guide the choice of the functional monomer for the preparation of MIPs. 4-Vinylpyridine and acrylic acid as optimal monomers were employed to synthesize MIPs for the

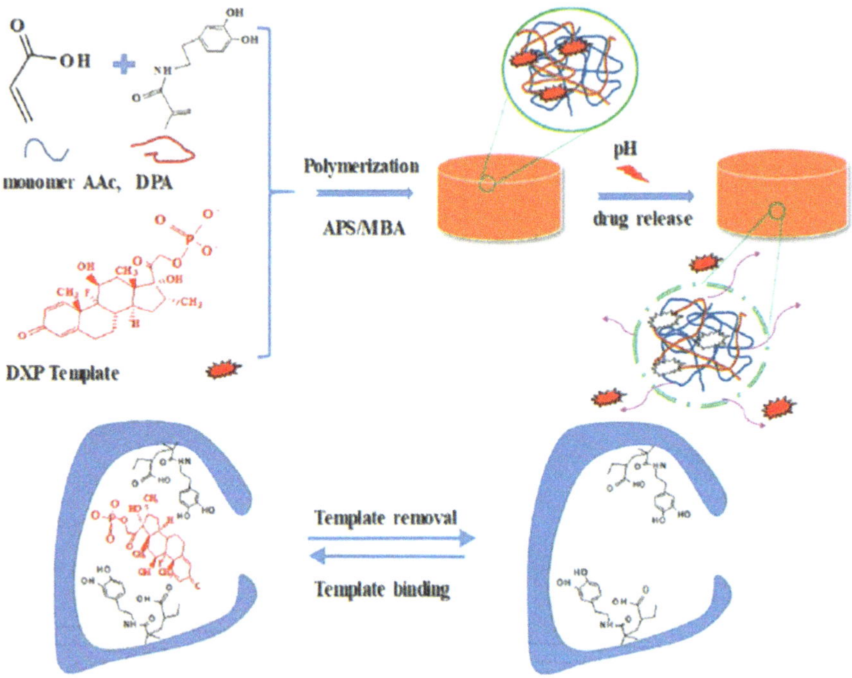

Fig. 4.2 The schematic diagram of the preparation of hydrogels. (Reproduced with permission of Ref. [12])

controlled release 5-fluorouracil. The similar release results are obtained, where the MIPs based on 4-vinylpyridine release 90% of 5-fluorouracil at pH 5.8, and the MIPs based on acrylic acid is 80% [14].

Besides the functional monomer, the proper choice of cross-linker is also important for the preparation of MIPs. When the template molecules are removed from the imprinting cavities, the integrity of polymer network structure depends on the cross-linker. Two traditional cross-linkers in molecular imprinting, ethylene glycol dimethacrylate and trimethylolpropane trimethacrylate were compared for the development of paclitaxel delivery system. The apparent release difference was observed for MIPs fabricated with the two cross-linkers. MIPs based on trimethylolpropane trimethacrylate demonstrated prolonged drug delivery ability, where the highest cumulative release was 85% in 50 h [15]. In another report, the same cross-linkers were also utilized to synthesize MIPs for doxorubicin, the results showed that the MIPs using ethylene glycol dimethacrylate as cross-linker were superior to that using trimethylolpropane trimethacrylate cross-linker, the adsorption capacity of the former was three times more than that of the latter [16].

Recently, some researchers have paid attention to exploit the novel reaction medium. Supercritical fluids have attracted many interests due to their physical and chemical properties. Supercritical carbon dioxide as one of the most extensively studied supercritical fluids is increasingly popular as polymerization reaction medium due to prominent physical property such as high diffusivity, high density, and low viscosity [17]. Polymer produced in such medium brings many advantages such as no organic solvents residues, omitted purification, and drying steps. An additional advantage in molecular imprinting is that supercritical carbon dioxide is benefit to interaction between the template and the monomers due to its apolar and aprotic property, leading to MIPs with higher affinity [18]. Casimiro and coworkers [19] developed a pH-sensitive MIPs for ibuprofen delivery. As reported in many previous papers, 2-(dimethylamino) ethyl methacrylate and ethylene glycol dimethacrylate were used as functional monomer and cross-linker, respectively. In order to enhance the biocompatibility, authors reduced the amount of cross-linker to produce the low cross-linked (20.2 wt%) MIPs in supercritical carbon dioxide medium. The release experiments verified that the preparation of low cross-linked MIPs in such medium was a promising strategy. Recently, a pH-responsive MIPs based on itaconic acid as functional monomer was also prepared in supercritical carbon dioxide medium. The resulted MIPs for metronidazole were investigated as a potential biocompatible drug delivery system [20].

Traditional hydrogels were usually prepared under photoinitiation polymerization at room temperature or thermal initiation polymerization at elevated temperature, which only provided hydrogels with small pore sizes. Macropore can facilitate the mass transfer during binding and release. Recently, a pH-responsive imprinted hydrogel having macropores have been prepared at subzero temperature, which is also called cryogels. *N*-Methacryloyl-L-glutamic acid (MAGA) as an acidic responsive monomer and (ethylene glycol) diacrylate as cross-linker was copolymerized and yielded a pH-responsive cryogel. The materials displayed a biphasic delivery behavior. The release results demonstrate a significantly enhanced rate for Doxorubicin in acidic medium. The release process is dominated by diffusion and erosion rate [21].

Unlike the traditional MIPs preparation technique, one-step molecular imprinting process avoiding the template elution is advantageous in drug delivery, where the template molecule encapsulated in MIPs is directly used to the following release. Such technique has been utilized to synthesize MIPs for the release of paclitaxel [22], and propranolol [23]. The pH-responsive MIPs release system has also been developed for the controlled release and delivery doxorubin [2] and vinblastine [24].

As previous statement, most of MIPs were prepared by co-polymerizing of functional monomers and cross-linkers in the presence of template molecule or its structural analogue, where large amount of organic solvents was used in the reaction and the template elution procedure. Ghaedi and coworker [25] reported an eco-friendly, less hazardous MIPs synthesis procedure for efficient delivery of riboflavin, where chitosan was utilized owing to their attractive nature and impressive advantages such as cheap, biodegradable, nontoxic, and biocompatibility, and simultaneously acted as functional monomer and cross-linker.

Although the specific loading and releasing drug could be achieved in theory, these MIPs particles loaded drug probably caused damage to healthy organs due to the lack of the specific recognition to organs. Thus, an external control to MIPs should be performed to avoid the side effects of the drug on healthy organs.

Magnetic materials have been reported in various fields. In controlled release system, drug can be conveniently positioned to target regional by an externally placed magnet. Therefore, magnetic Fe_3O_4 particles are undoubtful the ideal candidate due to easy preparation, excellent biocompatibility, and convenient operation by applying appropriate magnetic field. He and coworkers [26] combined the magnetic nanomaterials with MIPs to construct a water compatible magnetic MIPs release system for epirubicin. The resulted MIPs not only have magnetically susceptible characteristic, but also have pH-sensitive, highly selective recognition characteristics to the template molecule. In addition, a computational modeling approach was employed to facilitate the choice of functional monomer, and thus quickened the preparation process. A complex between methacrylic acid and epirubicin can formed by hydrogen bonds, van der Waals, where methacrylic acid simultaneously acted as a hydrogen bond donor and acceptor. The similar interaction was observed for another acidic monomer, acrylic acid. A typical basic functional, 4-vinylpyridine interacted with epirubicin by only van der Waals, where hydrogen bond interaction was not observed. The MIPs prepared using methacrylic acid showed the better loading and release performance than that of using methacrylamide. The proposed system can sustainably release all target molecule bonded by MIPs. Hydrogen bonding is predominant in the imprinting and release process. When a hollow magnetic particle is coated by a MIPs layer, the hollow magnetic MIPs particles delivery system is obtained. Recently, methacrylic acid and ethylene glycol dimethacrylate were used as functional monomer and cross-linker, respectively. The hollow Fe_3O_4 particles were first coated with methacryloxy propyltri methoxy silane and then reacted with methacrylic acid and ethylene glycol dimethacrylate in the presence of silybin via reverse atom radical transfer polymerization. The resulted hollow magnetic MIPs can maintain the release of silybin for 36 h in pH 2.0 medium at physiological temperature [27]. Graphene and graphene oxide have been widely reported in many fields due to their unique properties such as controllable electronic properties, multiple oxygen moieties and hydrophilicity. MIPs have been combined with such materials to improve the selectivity and the binding kinetic. Magnetic MIPs@graphene oxide for the release of rivastigmine was also prepared by grafting the copolymer of acrylamido-2-methyl-1-propanesulfonic acid and ethylene glycol diacrylate to the surface of acrylate functionalized Fe_3O_4 nanoparticles and graphene oxide. The resulted magnetic MIPs@graphene oxide exhibited the pH dependent release behavior in the range of 2.2–9.0 [28].

Besides the utilization of acidic or ionic functional monomer, the breaking of sulfur-sulfur bond can occur in the presence of containing sulfhydryl groups compound such as glutathione in an acidic pH medium. Thus, MIPs produced using the functional monomer containing sulfur–sulfur bond can display a double stimulus responsive. A double stimulus responsive drug delivery system for doxorubicin was developed by grafting MIPs to mesoporous silica nanoparticles. The loading amount and loading efficiency are 10.5 ± 0.2 wt% and $70 \pm 8\%$, respectively. High concentration glutathione and acidic medium can accelerate the release of doxorubicin [29].

Enantioselective-controlled delivery is important for chiral drugs. MIPs based on non-chiral or chiral functional monomers have been widely exploited for enantioselective-controlled release systems. However, MIPs based on non-chiral functional monomers exhibit a limited stereoselective release ability due to the non-defined recognition site [31, 32]. MIPs based on (S)-omeprazole was prepared by using a multifunctional chiral cinchona as functional monomer and ethylene glycol dimethacrylate as cross-linker via suspension polymerization. A pH-responsive hydrogel for the release of (S)-omeprazole was developed by combining the resulting MIPs with a poly (hydroxyethyl methacrylate) barrier and polycaprolactone-triol blend [33].

Unlike low-molar mass, protein possesses large molecular size, sophisticated structure, and various chemical groups, moreover is easily affected and denatured by surrounding conditions. Although many researches about protein imprinted polymers have been reported, it is still a huge challenge in the field of molecular imprinting. Recently, a potential oral drug delivery system for insulin was developed by using N,N-methylene-bisacrylamide as cross-linker, methacrylic acid and N-hydroxyethyl acrylamide as functional monomers. By comparing with NIPs, the resulted MIPs displayed not only a higher adsorption capacity, but also the enhanced release of insulin in solution at pH 7.4. In vivo studies on diabetic Wistar rats demonstrated that these MIPs nanoparticles were potential for oral insulin delivery [4].

4.2 Temperature-Sensitive Hydrogel-Based Molecularly Imprinted Polymers

Temperature-sensitive MIPs can change physical state such as swelling and shrinking under the stimulation of surrounding temperature, which can easily control the adsorption and release behavior, simultaneously avoid the slow mass transfer and incomplete release that encountered in traditional MIPs. Methacrylic acid, 2-(dimethylamino) ethyl methacrylate and N-isopropylacrylamide can responsive the external thermal stimulation, and are usually employed to produce the thermo-sensitive materials [34–36]. Among these monomers, N-isopropylacrylamide is the most well known as thermo-responsive monomer due to lower critical solution temperature. Most of temperature-sensitive MIPs are prepared using such monomer. Besides being responsible for the physical change, N-isopropylacrylamide with amide group can usually act as functional monomer to interact with template via hydrogen bonds and hydrophobic interaction. A computational research shows that N-isopropylacrylamide can interact with 5-fluorouracil via hydrogen bond, and the binding energy is about 246.50 kJ mol^{-1} [37]. In another report, Chen and coworkers combined the advantage of the magnetic particles with the thermal responsive function of the thermal MIPs. The thermal and magnetism bi-responsive MIPs for release of 5-fluorouracil were prepared by using N-isopropylacrylamide as a functional monomer and *N*,*N*′-methylene bisacrylamide

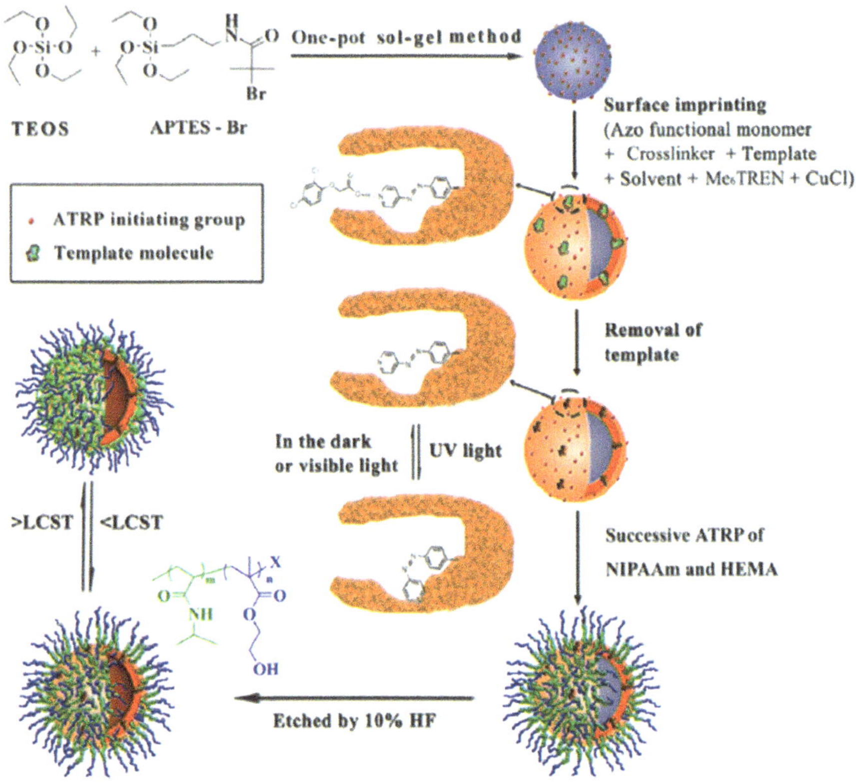

Fig. 4.3 Schematic illustration for the preparation of double responsive hydrophilic MIP microparticles. (Reproduced with permission of Ref. [40])

as a cross-linker. Authors pointed that the interaction energy between the recognition sites and 5-fluorouracil was $-112.24\ \text{kJ mol}^{-1}$, where hydrogen bond, a hydrophobic interaction and Van der Waals forces played important roles [38].

The thermal and magnetism bi-responsive MIPs for the release of curcumin were prepared by grafting MIPs layer consisted of acryl β-cyclodextrin and *N*-isopropylacrylamide onto the surface of Fe_3O_4 [39]. When the temperature-sensitive functional monomer and light-sensitive functional monomer were simultaneously contained into MIPs, which can yield a double responsive MIPs. A double responsive hydrophilic hollow MIPs microparticles for the controlled release 2,4-dichlorophenoxyacetic acid was described. In aqueous media, the resulted MIPs displayed pronounced light- and temperature-controlled template release behavior [40] (Fig. 4.3).

Although MIPs hydrogels have been widely demonstrated for the controlled release various drugs, MIPs for polar drugs are still in its initial stage. Inverse pickering emulsion polymerization approach is capable of synthesizing hydrophilic MIPs directly in water [41]. A micro-spherical imprinted hydrogel was prepared via inverse pickering emulsion for the controlled release of adenosine 5′-monophosphate [42].

4.3 Cyclodextrins-Based Molecularly Imprinted Polymers

Due to unique amphiphilic with a hydrophilic external and a hydrophobic cavity, cyclodextrins can form inclusion complexes with various organic molecule through host-guest interactions or non-covalent, thus modifying their solubility, physicochemical property and physiological stability. It has been widely used to deliver various active pharmaceutical ingredients in the pharmaceutical industry [43–45] (Fig. 4.4).

During the imprinting process, the interaction between cyclodextrin and template molecule can be positioned by cross-linker. The specific molecular cavity of cyclodextrin endows MIPs with better selectivity [46] (Fig. 4.5).

An affinity-based cyclodextrin delivery platform has demonstrated a prolonged release of antibiotics beyond 45 days [47]. The drug delivery hydrogel based on cyclodextrins can effectively prevent staphylococcus aureus mesh infections [48].

Although MIPs system has demonstrated the potential application in the release of medicines and loading [49, 50], the interaction between template and functional monomer is usually weakened by water molecules, which leads to the poor binding and release property in aqueous medium. In order to enhance MIPs property of the release and loading, cyclodextrin has been introduced into the field of MIPs, where

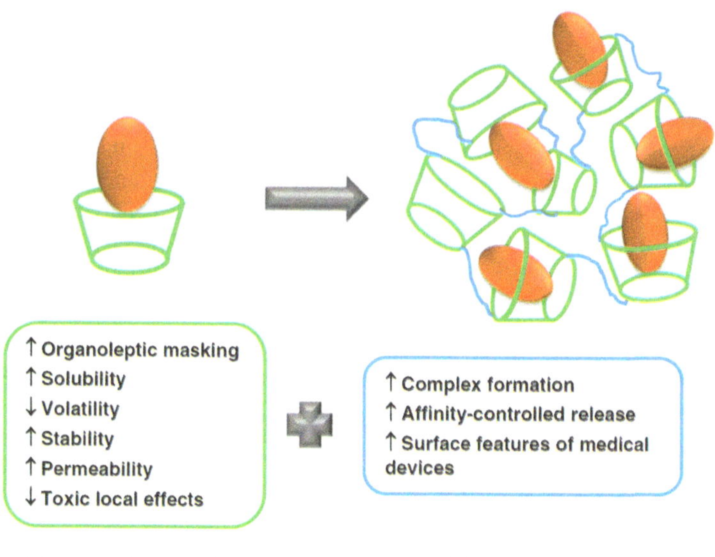

Fig. 4.4 Advantages by inclusion complex formation with cyclodextrins and by cross-linked cyclodextrin networks. (Reproduced with permission of Ref. [45])

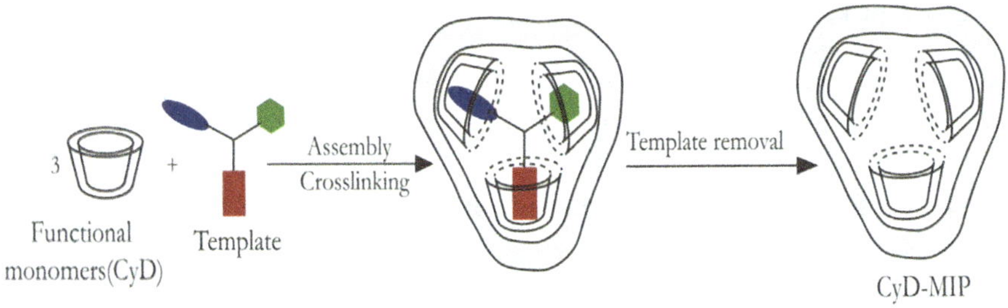

Fig. 4.5 Interaction between cyclodextrin and template during molecular imprinting. (Reproduced with permission of Ref. [46])

cyclodextrins with active hydroxyl groups are easily linked with another by different types of active molecules or they can be functionalized with polymerizable monomer and used as a functional monomer. Zhang and co-workers reported the preparation of chitosan based MIPs via surface imprinting, where the interaction between β-cyclodextrin grafted to the surface of chitosan and sinomenine hydrochloride (template molecule) was oriented by polymerization of methacrylic acid and ethylene glycol dimethacrylate at 80 °C for 12 h. The MIPs with maximum binding capacity (55.9 mg g^{-1}) was obtained under the optimal composition ratio (6 mmol of functional monomer and 20 mmol cross-linker). In vitro drug release studies, the accumulative release amount of MIPs was up to 78% within 24 h [51]. The complexes of β-cyclodextrin with steroids was facially positioned by using toluene 2,4-diisocyanate at 65 °C via one-step procedure, where two and three molecules of β-cyclodextrin are simultaneously interacted with large steroids. These MIPs selectively adsorbed with cholesterol or stigmasterol in aqueous medium, however, steroids that can only interact with one cyclodextrin is failed to be imprinted by this approach [52]. MIPs based on polymerizable monomer functionalized cyclodextrin such as acryolyl-⟨-cyclodextrin and acryloyl-(6-O-α-D-glucosyl)-β-cyclodextrin were also explored for the large target molecules (some dipeptides, cefazolin, vancomycin, and phenethicillin), where one molecule can simultaneously interact with several cyclodextrin units. In such procedure, additional functional monomers and a cross-linker monomer are necessary to positioned the spatially arrangement of the cyclodextrin. Compared with NIPs, the resultant cyclodextrin-based MIPs adsorbed twice the amount of drug [53, 54]. Bisacryloyl-β-cyclodextrin monomers and ionic 2-acryloylamido-2,2′-dimethylpropane sulfonic acid monomers were used to imprint amphiphilic phenylalanine, where functional monomer and template molecule can interact via the hydrophobic and electrostatic interactions. The resulted MIPs demonstrated the excellent recognition property [55]. Recently, the β-cyclodextrin-based MIPs for the release of atropine was prepared by using methacrylic acid functionalized beta-cyclodextrin as functional monomer and trimethylolpropane trimethacrylate as cross-linker via thermally-initiated precipitation polymerization technique. MIPs prepare using methacrylic acid functionalized beta-cyclodextrin demonstrated the more binding template molecules than that using MAA. MIPs displayed not only a high ATP loading capacity, but also a longer

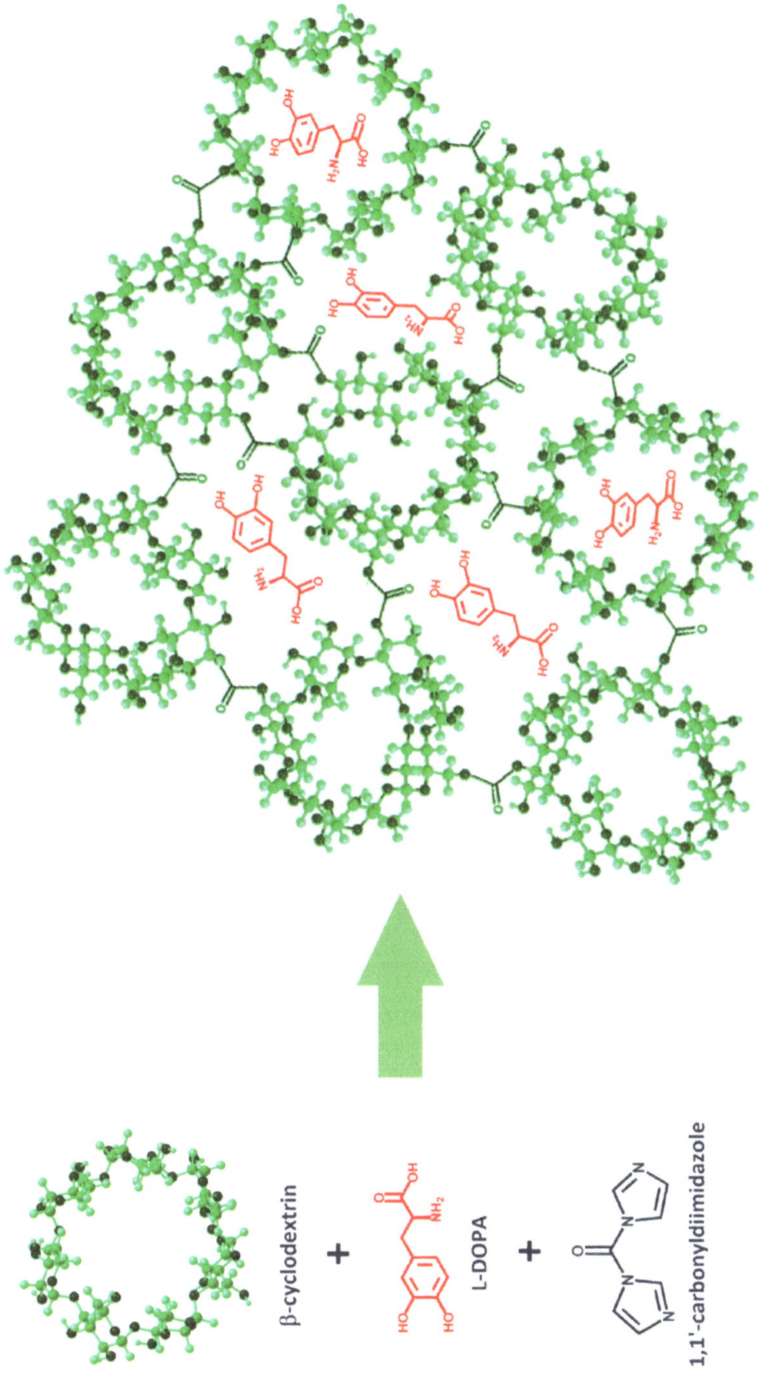

Fig. 4.6 Schematic representation of the preparation of L-DOPA molecularly imprinted CD-NS. (Reproduced with permission of Ref. [60])

release activity that NIPs. Moreover, the release rate of MIPs increased when pH value decreased in the range from 7.4 to 1.5. [56]. To further enhance the effectiveness of loading and delivery, molecularly imprinted supramolecular hydrogels based on cyclodextrin-based macromonomers were prepared at room temperature. The resulted macromolecule hydrogels displayed the sustainable release over a period of 35 days [57]. N-(hydroxymethyl) acrylamide functionalized γ-cyclodextrin as functional monomer and N,N-methylene (bisacrylamide) as cross-linker were used to prepare MIPs for loading and release of the hydrophilic propranolol and the hydrophobic triamcinolone acetonide. During the imprinting and recognition process, triamcinolone acetonide and γ-cyclodextrin can form complexes at the molar ratio of 1:1, and the resulted MIPs demonstrated a significantly higher loading capacity for triamcinolone acetonide. Such MIPs displayed the enhanced swelling degree in the mixture solvent of ethanol and water, in which triamcinolone acetonide is easily dissolved. However, the hydrophilic characteristic of PR resulted in the weaker interaction with γ-cyclodextrin, and thus MIPs failed to improve the loading and release of propranolol [58, 59].

Trotta et al. was the first to report a paper on MIP cyclodextrin-NSs applied to the study of an oral formulation for the delivery of L-DOPA for the treatment of neurodegenerative disease. MIP-NSs were synthesized through condensation polymerization by reacting b-cyclodextrin with different amounts of 1,10-carbonyldiimidazole in the presence of a variable amount of drug. Cyclodextrin cavity allowed the formation of the inclusion compound with unstable and reactive L-DOPA, shielding its reactive groups and preventing its degradation [60] (Fig. 4.6).

References

1. Malakooti N, Alexander C, Alvarez-Lorenzo C (2015) Imprinted contact lenses for sustained release of polymyxin B and related antimicrobial peptides. J Pharm Sci 104:3386–3394
2. Kaamyabi S, Habibi D, Amini MM (2016) Preparation and characterization of the pH and thermosensitive magnetic molecular imprinted nanoparticle polymer for the cancer drug delivery. Bioorg Med Chem Lett 26:2349–2354
3. Zhu Y, Yang L, Huang D, Zhu Q (2017) Molecularly imprinted nanoparticles and their releasing properties, bio-distribution as drug carriers. Asian J Pharm Sci 12:172–178
4. Paul PK, Treetong A, Suedee R (2017) Biomimetic insulin-imprinted polymer nanoparticles as a potential oral drug delivery system. Acta Pharm 67:149–168
5. Allender CJ, Richardson C, Woodhouse B, Heard CM, Brain KR (2000) Pharmaceutical applications for molecularly imprinted polymers. Int J Pharm 195:39–43
6. Alvarez-Lorenzo C, Concheiro A (2006) Molecularly imprinted materials as advanced excipients for drug delivery systems. Biotechnol Annu Rev 12:225–268
7. Sellergren B, Allender CJ (2005) Molecularly imprinted polymers: a bridge to advanced drug delivery. Adv Drug Deliv Rev 57:1733–1741
8. Abdouss M, Asadi E, Azodi-Deilami S, Beik-mohammadi N, Aslanzadeh SA (2011) Development and characterization of molecularly imprinted polymers for controlled release of citalopram. J Mater Sci Mater Med 22:2273–2281

9. Wang C, Javadi A, Ghaffari M, Gong S (2010) A pH-sensitive molecularly imprinted nanospheres/hydrogel composite as a coating for implantable biosensors. Biomaterials 31:4944–4951
10. Mao C, Xie X, Liu X, Cui Z, Yang X, Yeung KWK, Pan H, Chu PK, Wu S (2017) The controlled drug release by pH-sensitive molecularly imprinted nanospheres for enhanced antibacterial activity. Mater Sci Eng C Mater Biol App 77:84–91
11. Yan Q, Liu L, Wang T, Wang H (2017) A pH-responsive hydrogel system based on cellulose and dopamine with controlled hydrophobic drug delivery ability and long-term bacteriostatic property. Colloid Polym Sci 297:705–717
12. Wei P, Song R, Chen C, Li Z, Zhu Z, Li S (2019) A pH-responsive molecularly imprinted hydrogel for dexamethasone release. J Inorg Organomet Polym Mater 29:659–666
13. Mohajeri SA, Malaekeh-Nikouei B, Sadegh H (2012) Development of a pH-responsive imprinted polymer for diclofenac and study of its binding properties in organic and aqueous media. Drug Dev Ind Pharm 38:616–622
14. Talavat L, Güner A (2019) Thermodynamic computational calculations for preparation 5-fluorouracilmagnetic moleculary imprinted polymers and their application in controlled drug release. Inorg Chem Commun 103:119–127
15. Cegłowski M, Kurczewska J, Ruszkowski P, Schroeder G (2019) Application of paclitaxel-imprinted microparticles obtained using two different cross-linkers for prolonged drug delivery. Eur Polym J 118:328–336
16. Cegłowski M, Kurczewska J, Ruszkowski P, Liberska J, Schroeder G (2019) The influence of cross-linking agent onto adsorption properties, release behavior and cytotoxicity of doxorubicin-imprinted microparticles. Colloids Surf B Biointerfaces 182:110379
17. Cooper AI (2001) Recent developments in materials synthesis and processing using supercritical CO2. Adv Mater 13:1111–1114
18. Duarte ARC, Casimiro T, Aguiar-Ricardo A, Simplício AL, Duarte CMM (2006) Supercritical fluid polymerisation and impregnation of molecularly imprinted polymers for drug delivery. J Supercrit Fluids 39:102–106
19. da Silva MS, Viveiros R, Morgado PI, Aguiar-Ricardo A, Correia IJ, Casimiro T (2011) Development of 2-(dimethylamino)ethyl methacrylate-based molecular recognition devices for controlled drug delivery using supercritical fluid technology. Int J Pharm 416:61–68
20. Marcelo G, Ferreira IC, Viveiros R, Casimiro T (2018) Development of itaconic acid-based molecular imprinted polymers using supercritical fluid technology for pH-triggered drug delivery. Int J Pharm 542:125–131
21. Çetin K, Alkan H, Bereli N, Denizli A (2017) Molecularly imprinted cryogel as a pH-responsive delivery system for doxorubicin. J Macromol Sci Part A 8:502–508
22. Esfandyari-Manesh M, Darvishi B, Ishkuh FA, Shahmoradi E, Mohammadi A, Javanbakht M, Dinarvand R, Atyabi F (2016) Paclitaxel molecularly imprinted polymer-PEG-folate nanoparticles for targeting anticancer delivery: characterization and cellular cytotoxicity. Mater Sci Eng C Mater Biol Appl 62:626–633
23. Barde LN, Ghule MM, Roy AA, Mathur VB, Shivhare UD (2013) Development of molecularly imprinted polymer as sustain release drug carrier for propranolol HCL. Drug Dev Ind Pharm 39:1247–1253
24. Zhu Y, Liu R, Huang H, Zhu Q (2019) Vinblastine-loaded nanoparticles with enhanced tumor-targeting efficiency and decreasing toxicity: developed by one-step molecular imprinting process. Mol Pharm 16:2675–2689
25. Mokhtari P, Ghaedi M (2019) Water compatible molecularly imprinted polymer for controlled release of riboflavin as drug delivery system. Eur Polym J 118:614–618
26. Dramou P, Zuo P, He H, Pham-Huy LA, Zou W, Xiao D, Pham-Huy C, Ndorbor T (2013) Anticancer loading and controlled release of novel water-compatible magnetic nanomaterials as drug delivery agents, coupled to a computational modeling approach. J Mater Chem B 1:4099–4109

27. Ji K, Luo X, He L, Liao S, Hu L, Han J, Chen C, Liu Y, Tan N (2020) Preparation of hollow magnetic molecularly imprinted polymer and its application in silybin recognition and controlled release. J Pharm Biomed Anal 180:113036
28. Hemmati K, Sahraei R, Ghaemy M (2016) Synthesis and characterization of a novel magnetic molecularly imprinted polymer with incorporated graphene oxide for drug delivery. Polymer 101:257–268
29. Zhang K, Guan X, Qiu Y, Wang D, Zhang X, Zhang H (2016) A pH/glutathione double responsive drug delivery system using molecular imprint technique for drug loading. Appl Surf Sci 389:1208–1213
30. Suedee R, Srichana T, Martin GP (2000) Evaluation of matrices containing molecularly imprinted polymers in the enantioselective-controlled delivery of beta-blockers. J Control Release 66:135–147
31. Suedee R, Srichana T, Rattananont T (2002) Enantioselective release of controlled delivery granules based on molecularly imprinted polymers. Drug Deliv 9:19–30
32. Bodhibukkana C, Srichana T, Kaewnopparat S, Tangthong N, Bouking P, Martin GP, Suedee R (2006) Composite membrane of bacterially-derived cellulose and molecularly imprinted polymer for use as a transdermal enantioselective controlled-release system of racemic propranolol. J Control Release 113:43–56
33. Suedee R, Jantarat C, Lindner W, Viernstein H, Songkro S, Srichana T (2010) Development of a pH-responsive drug delivery system for enantioselective-controlled delivery of racemic drugs. J Control Release 142:122–131
34. Chen F, Jiang XP, Kuang TR, Chang LQ, Fu DJ, Zhong MQ (2015) Polyelectrolyte/mesoporous silica hybrid materials for the high performance multiple-detection of pH value and temperature. Polym Chem 6:3529–3536
35. Chen F, Jiang XP, Kuang TR, Chang LQ, Zhong MQ (2015) Effect of nanoporous structure and polymer brushes on the ionic conductivity of poly(methacrylic acid)/anode aluminum oxide hybrid membranes. RSC Adv 5:70204–70210
36. Zhang C, Jia X, Wang Y, Zhang M, Yang S, Guo J (2014) Thermosensitive molecularly imprinted hydrogel cross-linked with N-malely chitosan for the recognition and separation of BSA. J Sep Sci 37:419–426
37. Zhang L, Chen L, Zhang H, Yang Y, Liu X (2017) Recognition of 5-fluorouracil by thermosensitive magnetic surface molecularly imprinted microspheres designed using a computational approach. J Appl Polym Sci 134:45468
38. Li L, Chen L, Zhang H, Yang Y, Liu X, Chen Y (2016) Temperature and magnetism bi-responsive molecularly imprinted polymers: preparation, adsorption mechanism and properties as drug delivery system for sustained release of 5-fluorouracil. Mater Sci Eng C Mater Biol Appl 61:158–168
39. Sedghi R, Yassari M, Heidari B (2018) Thermo-responsive molecularly imprinted polymer containing magnetic nanoparticles: synthesis, characterization and adsorption properties for curcumin. Colloids Surf B Biointerfaces 162:154–162
40. Li C, Ma Y, Niu H, Zhang H (2015) Hydrophilic hollow molecularly imprinted polymer microparticles with photo- and thermoresponsive template binding and release properties in aqueous media. ACS Appl Mater Interfaces 7:27340–27350
41. Shen X, Ye L (2011) Molecular imprinting in Pickering emulsions: a new insight into molecular recognition in water. Chem Commun (Camb) 47:10359–10361
42. Ayari MG, Kadhirvel P, Favetta P, Plano B, Dejous C, Carbonnier B, Agrofoglio LA (2019) Synthesis of imprinted hydrogel microbeads by inverse Pickering emulsion to controlled release of adenosine 5′-monophosphate. Mater Sci Eng C Mater Biol App 101:254–263
43. Loftsson T, Brewster ME (1996) Pharmaceutical applications of cyclodextrins. 1. Drug solubilization and stabilization. J Pharm Sci 85:1017–1025
44. Loftsson T, Duchêne D (2007) Cyclodextrins and their pharmaceutical applications. Int J Pharm 329:1–11

45. Concheiro A, Alvarez-Lorenzo C (2013) Chemically cross-linked and grafted cyclodextrin hydrogels: from nanostructures to drug-eluting medical devices. Adv Drug Deliv Rev 65:1188–1203
46. Adeoye O, Cabral-Marques H (2017) Cyclodextrin nanosystems in oral drug delivery: a mini review. Int J Pharm 531:521–531
47. Zhang H, Feng W, Li C, Tan T (2010) Investigation of the inclusions of puerarin and daidzin with beta-cyclodextrin by molecular dynamics simulation. J Phys Chem B 114:4876–4883
48. Henriksen NM, Fenley AT, Gilson MK (2015) Computational calorimetry: high-precision calculation of host-guest binding thermodynamics. J Chem Theory Comput 11:4377–4394
49. Cunliffe D, Kirby A, Alexander C (2005) Molecularly imprinted drug delivery systems. Adv Drug Deliv Rev 57:1836–1853
50. Alvarez-Lorenzo C, Yañez F, Barreiro-Iglesias R, Concheiro A (2006) Imprinted soft contact lenses as norfloxacin delivery systems. J Control Release 113:236–244
51. Chen H, Zhang W, Yang N, Chen C, Zhang M (2018) Chitosan-based surface molecularly imprinted polymer microspheres for sustained release of sinomenine hydrochloride in aqueous media. Appl Biochem Biotechnol 185:370–384
52. Hishiya T, Shibata M, Kakazu M, Asanuma H, Komiyama M (1999) Molecularly imprinted cyclodextrins as selective receptors for steroids. Macromolecules 32:2265–2269
53. Hishiya T, Asanuma H, Komiyama M (2002) Spectroscopic anatomy of molecular-imprinting of cyclodextrin. Evidence for preferential formation of ordered cyclodextrin assemblies. J Am Chem Soc 124:570–575
54. Asanuma H, Akiyama T, Kajiya K, Hishiya T, Komiyama M (2001) Molecular imprinting of cyclodextrin in water for the recognition of nanometer-scaled guests. Anal Chim Acta 435:25–33
55. Piletsky SA, Andersson HS, Nicholls IA (1999) Combined hydrophobic and electrostatic interaction-based recognition in molecularly imprinted polymers. Macromolecules 32:633–636
56. He Y, Zeng S, Abd El-Aty AM, Hacımüftüoğlu A, Kalekristos Yohannes W, Khan M, She Y (2020) Development of water-compatible molecularly imprinted polymers based on functionalized β-cyclodextrin for controlled release of atropine. Polymers (Basel) 12:130
57. Juric D, Rohner NA, von Recum HA (2019) Molecular imprinting of cyclodextrin supramolecular hydrogels improves drug loading and delivery. Macromol Biosci 19:e1800246
58. Caldera F, Tannous M, Cavalli R, Zanetti M, Trotta F (2017) Evolution of cyclodextrin nanosponges. Int J Pharm 531:470–479
59. Siemoneit U, Schmitt C, Alvarez-Lorenzo C, Luzardo A, Otero-Espinar F, Concheiro A, Blanco-Méndez J (2006) Acrylic/cyclodextrin hydrogels with enhanced drug loading and sustained release capability. Int J Pharm 312:66–74
60. Trotta F, Caldera F, Cavalli R, Soster M, Riedo C, Biasizzo M, Barretta GU, Balzano F, Brunella V (2016) Molecularly imprinted cyclodextrin nanosponges for the controlled delivery of L-DOPA: perspectives. Expert Opin Drug Deliv 13:1671–1680

Chapter 5
Stimuli Responsive Imprinted DDS

Li-Ping Zhang and Zhaosheng Liu

The molecularly imprinted polymers (MIPs) with molecular recognition abilities are stable polymers and could resist the change of outside condition, such as the temperature, pH, organic solvents, and pressure [1]. Thus, the MIPs could simulate the interactions between antibody and antigen, enzymes and receptors without the stability limitations. MIPs have been used for a number of applications, e.g., chromatographic stationary phases [2, 3], solid-phase extraction (SPE) [4], catalysis [5], antibodies and enzyme mimics [6], affinity, and sensing materials [7]. The MIPs also present the enormous potential in several drug delivery systems (DDS), which could control the release of drug by the flexible polymer network that cross-linked with functional group [8–10].

The advantages of MIPs in DDS attribute to the abilities of controlling release of drug and the higher loading capacity [11]. When placed in the medium after drug loading, the template will diffuse from the MIP through the mechanism of adsorption–desorption [12]. When drug loading and the release process is in a state of balance, the concentration of template in the MIP will reach a stable level, so as to realize the zero-order release [9, 10]. In addition, the affinity between template and functional groups in MIPs is relative strong, which makes MIPs emerge higher drug loading capacity than traditional non-imprinted polymers. What's more, the MIPs acted as DDS could keep the plasma concentration below toxic levels since the drug has a narrow therapeutic window.

The smart responsive polymers are a class of material that could respond to the specific environment stimuli (temperature, light, pH, electric field, ionic strength, etc.) and display the change in properties involved in solubilities, structures, and

L.-P. Zhang
School of Basic Medical Sciences, Henan University of Science and Technology, Luoyang, China

Z. Liu (✉)
College of Pharmacy, Tianjin Medical University, Tianjin, China

Z. Liu et al. (eds.), *Molecularly Imprinted Polymers as Advanced Drug Delivery Systems*, https://doi.org/10.1007/978-981-16-0227-6_5

shapes [13]. The novel smart responsive MIPs have been developed by combining the smart responsive polymers and MIPs, of which the molecular chain structure, swelling degree, surface properties, solubility, and imprinting site of them have changed under the stimulation of the external environment. It can be explained in the following aspects: when stimulated by the external environment, the imprinted polymer would change in the molecular microstructure, leading to the weakening or disappearance of the 3D structure of imprinted cavities, and the loss of the specific recognition ability of the polymer to the template molecule [14]. However, when eliminating the external conditions, the specific recognition performance of MIPs could resume. Therefore, the adsorption and elution of template in MIPs can be successfully controlled by the response of external stimuli such as temperature, pH and solvent.

The combination of smart responsive polymers and MIPs may reveal some considerable advantages. For example, MIPs can provide higher loading capacity for specific molecules, the ability to respond to external stimuli, and the capability of the loading/release processes by modulating the affinity of the network to the templates. Therefore, the smart MIPs presented tremendous potential in several incurable diseases including cancer [15], Alzheimer's disease [16], Osteomyelitis [17], bacterial, and ocular infections [18, 19]. These systems respond to their environment exclusively, allowing drugs to be delivered to the target site, in response to the disease-related stimulus.

5.1 Thermo-Responsive Drug Delivery System and the Application of MIP

Thermosensitive drug delivery system is a drug preparation with temperature response, which can be obtained by combining the drugs, thermosensitive materials and appropriate drug carrier with a certain form. As a drug carrier, the thermosensitive complexes could regulate the upload and release of drugs through the change of temperature, so as to realize the controlled release of drug molecules, adjust the release rate of drugs in vivo, maintain the plasma concentration for a long time, and effectively improve the bioavailability of drugs [20]. Among them, the thermosensitive units play a decisive role, in which the conversion of hydrophilic and hydrophobic groups in the complex could cause internal structural change with the variation of ambient temperature.

So far, the thermosensitive materials that have been successfully prepared, including hydrogels, nanoparticles, liposomes, and so on. The poly(*N*-isopropylacrylamide) (PNIPAM), polyvinyl pyrrolidone (PVP), polyethylene glycol (PEG) and polyvinyl alcohol (PVA) are commonly used as the thermosensitive materials. Due to its low critical solution temperature (LCST) in the range of 25–32 °C, the (PNIPAM) is the most widely studied, i.e., close to the temperature of the human body [21]. As far, it has been widely studied in drug sustained release and other biological fields [22].

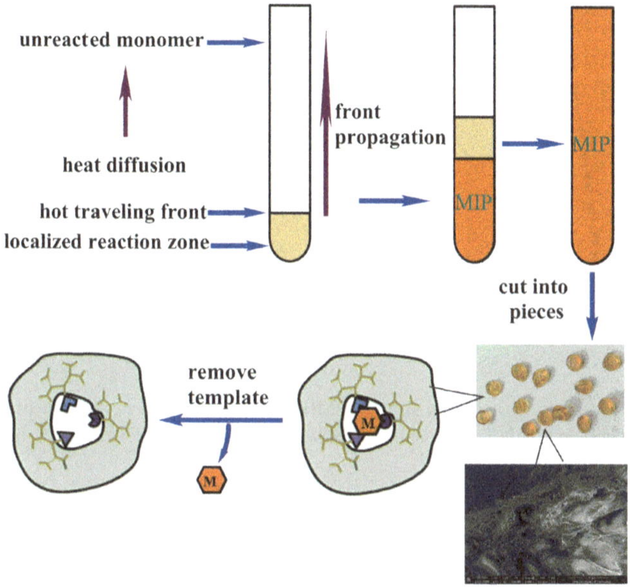

Fig. 5.1 Procedures of synthesizing MIP in frontal polymerization mode

The thermo-responsive MIP hydrogels are a kind of gels that could respond to the change of temperature and result in the swelling or shrinkage. The thermosensitive molecularly imprinted polymer (T-MIP) is prepared by the combination of molecular imprinting technology and thermosensitive materials. The obtained T-MIP possess the advantages of imprinting and thermo-sensibility at the same time, which not only improve the specific recognition ability of polymer to target, but also realize the controlled loading and release of templates under the change of external temperature [23, 24]. Up to now, there have been many ways to prepare the hydrogels, including bulk polymerization, sacrificial template method, post-crosslinking, and interpenetrating network copolymerization. By the thermosensitive functional monomers, they all response the temperature changes, other auxiliary monomers are added for binding and fixation of template. However, hydrogels-based imprinting is more difficult than rigid polymer-based imprinting.

All structural considerations controlling the architecture of polymer for hydroMIP preparation include the number and diversity of functional monomeric species, the interaction strength of template-functional monomer, as well as polymerization mode [25], which has been demonstrated to affect affinity of MIP and control the size of the imprinted cavities. Recently, frontal polymerization (FP) is found an alternative synthetic technique for polymer preparation (Fig. 5.1), which allows the conversion of monomer into polymer in a localized reaction zone [26]. Compared to batch polymerization (BP), FP is able to self-sustain and propagate throughout the monomeric mixture and has shorter reaction time, a simpler reaction route, and lower energy consumption. FPs can fabricate polymer with high

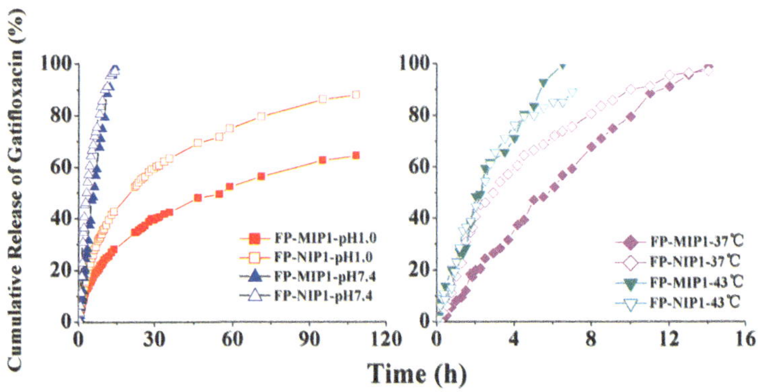

Fig. 5.2 Release profiles of GFLX from FP-MIP and FP-NIP under different pH (pH 1.0 HCl aqueous solution and pH 7.4 phosphate buffer) and temperature (37° and 43°) conditions. (Reproduced with permission of [24])

conversion rates at short reaction times due to their self-propagating nature. In addition, the hydrogel prepared by FP exhibited higher swelling ratio than that prepared by BP. Thus, FP provides an alternative method to prepare MIPs-based DDS.

FP has been successfully used to prepare hydroMIPs [24]. By this approach, the hydroMIPs against gatifloxacin (GFLX) was prepared using a mixture of acrylic acid, *N*-isopropylacrylamide, as well as *N*,*N*′-methylenebisacrylamide in 20 min. Higher affinity of the gatifloxacin-imprinted polymer made by FP than BP was found. From Fig. 5.2, an extremely slow release of GFLX on the hydroMIP is showed with the duration of over 120 h at pH 1.0. In addition, a slower release of GFLX on the hydroMIP system was observed than that on the hydroNIP.

As shown in pharmacokinetic profiles, the loaded FP-based hydroMIP showed the shortest T_{max} of 1.5 h. In contrast, T_{max} of the FP-based hydroNIP and BP-based hydroMIP was 3.0 and 2.0 h, respectively. However, the plasma concentration with a plateau region of slightly lower C_{max} between 2 and 10 h was found due to the controlled release of GFLX from the FP-based hydroMIP. Moreover, the highest value of AUC_{0-12} (220.9 h ng/mL) was observed for the FP-based hydroMIP. In contrast, the AUC_{0-12} value for the FP-based hydroNIP and BP-based hydroMIP was 56.7 and 131.1 h ng/mL, respectively.

5.2 pH-Sensitive Hydrogel-Based Molecularly Imprinted Polymers

Hydrogels are insoluble cross-linked macromolecular networks of polymer chains swollen in water. They contain over 99% water due to their hydrophilic absorbent nature. The disadvantage of hydrogels is relatively low mechanical strength, which can be overcome by either cross-linking or crystallisation. Hydrogels have become

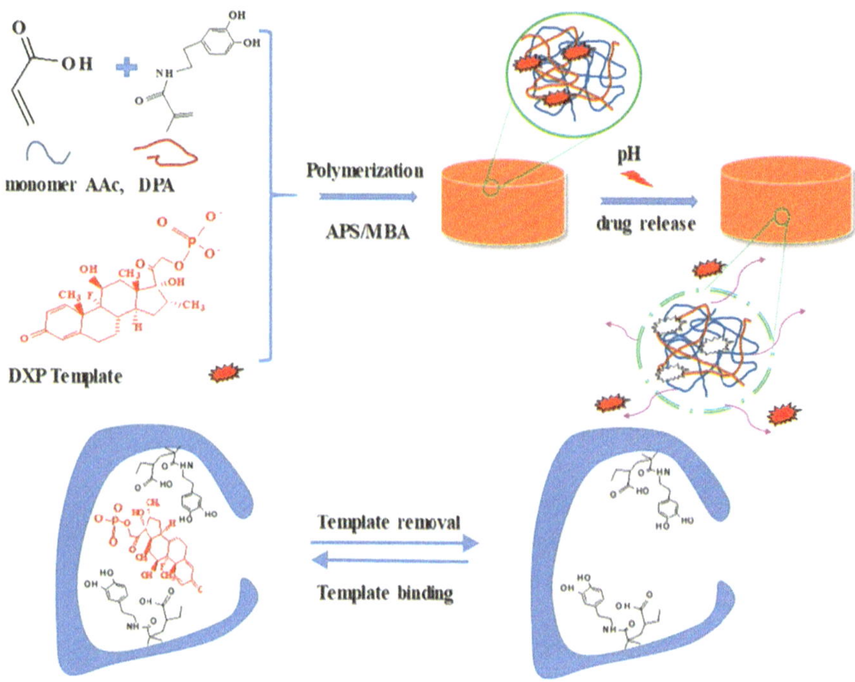

Fig. 5.3 Schematic diagram of the preparation of pH-responsive MIP hydrogel. (Reproduced with permission of [28])

excellent carriers for drugs release. For example, polyacrylamide (PAM) is particularly suitable for controlling the latter factors for effective design and synthesis. PAM hydrogels possess the necessary parameters to successfully produce an analyte-specific MIP [27], i.e., being very inert, offering hydrogen bonding capabilities, and being biocompatible, which is attributed to their ability to mimic natural tissue due to their high water content and their special surface properties. As a nitrogen containing member of the acrylate polymers, PAM is identified as a suitable imprinting matrix as it is cheap, water soluble, easily produced with attractive structural parameters. *N′,N′*-methylenebisacrylamide (MBAm), a dimer of AAm is usually used as the cross-linker [28].

To control the release of dexamethasone sodium phosphate (DXP), a new type of MIP hydrogel with pH-responsive properties was reported (Fig. 5.3). By precipitation polymerization, dopamine acrylamide (DPA) and acrylic acid (AAc) were used to prepare the MIP hydrogel. Higher drug loading of the resulting MIP hydrogel than NIP hydrogel can be achieved. Only 58% of the loaded drug released in 24 h from the MIP hydrogel. In contrast, 80% of the released drug was found for the NIP hydrogel. The MIP hydrogel can sustain such controlled slow release characteristics up to 7 days.

5.3 Solvent-Responsive Floating Molecularly Imprinted Polymers

To improve low bioavailability and subsequent therapeutic efficacy of the drug, gastric-resident and gastric-retentive drug delivery systems (GRDDS) is developed by retaining drug in stomach with a controlled release. Flotation was frequently studied as a powerful method to bypass the gastric emptying process. Preparing floating drug delivery system (FDDS) is expensive and hard because low-density dry solid systems are often made for the intragastric buoyancy. The floating behavior of the particles of liquid crystalline-molecularly imprinted polymer (LC-MIP) in the aqueous medium were first found on 2017 [29]. After that, the LC-MIP was regarded as a novel class of the gastric-retentive DDS based on molecular recognition, which will be discussed in the subsequent context.

5.3.1 Liquid Crystalline-Molecularly Imprinted Polymers for Gastroretentive Controlled Drug Release System

Liquid crystal elastomers (LCEs) are materials with a combination of the flexibility of loosely cross-linked polymer network as well as the anisotropy of liquid crystal. Thus, LCEs may alter form reversibly in response to a number of external stimulus (electrical or magnetic field) and environmental factors (temperature, light, or solvent). Up to date, LCEs can be used to generate many functional devices, e.g., non-linear optical devices, actuators and sensors, artificial muscles, and surface-feature patterning [30].

Recently, the concept of LCEs has been used in the field of MIP preparation. With the introduction of liquid crystalline (LC) monomers, the memory of template of the obtained LC-MIP can be preserved even at very low amount of cross-linker (5–10 mol%) [29]. Because of the interactions of mesogenic groups, conventional chemical cross-linking can be replaced by physical cross-linking. Thus, the LC-MIP reinforce the memory effect to template molecule with greater trapping capacity because of the decrease in embedding sites. Moreover, the property of shape memory was found for imprinted LCE, as shown in the application of artificial enzyme. The drawbacks of the demand of high percentage of cross-linker (80–90 mol%), such as poor site accessibility, for MIPs preparation can be overcome. In addition, the LC-MIP displayed phase transition behaviors.

Typically, LC-MIP can be prepared using 4-methyl phenyl dicyclohexyl ethylene (MPDE) (LC monomer), template molecule, methacrylic acid, and EDMA. In spite of smaller imprinting factors (2.80) for the LC-MIP at the cross-linking degree of 20.0% compared with the LC-free MIP (6.70) with 80.0% chemical cross-linker, outstanding floating properties of the LC-MIP and corresponding LC-NIP were found in water (Fig. 5.4). For instance, they floated directly to the upper surface when placed in water, and remained buoyant for at least 24 h. In contrast, LC-free

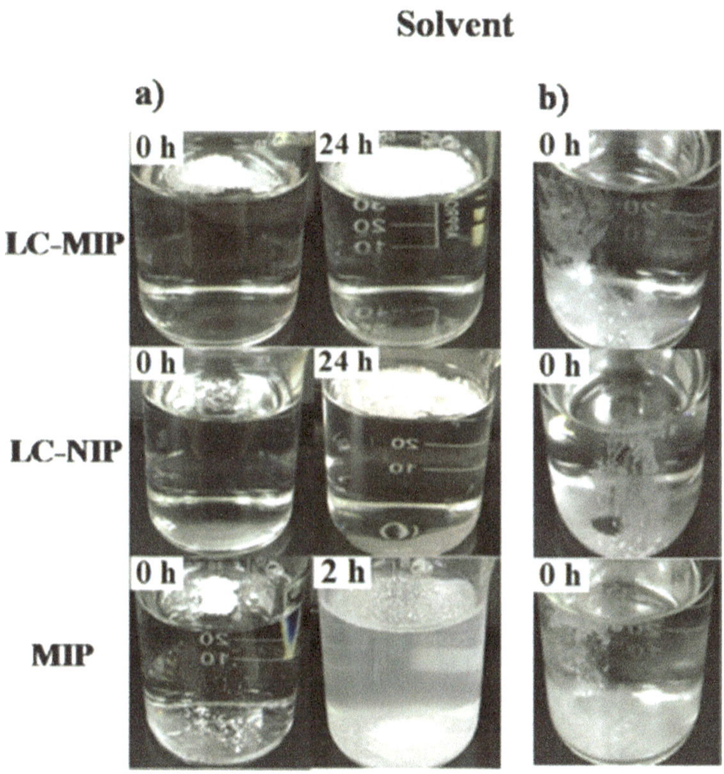

Fig. 5.4 Floating and solvent-sensitivity of LC-MIP, LC-NIP, and control LC-free MIP in aqueous medium (**a**) and organic solvent (**b**). (Reproduced with permission of [29])

MIP sank to the bottom. In contrast, LC-MIP and LC-NIP sank to the bottom when ACN or chloroform, a solvent of high density, was used as the medium.

The floating LC-MIP can give longer gastric residence time than the MIP that can not float (Fig. 5.5). This can be proved by rhodamin B (RB) adsorption on the LC-MIP and LC-free MIP. For example, after the immersion in water for 2–3 h, fluorescence was still 26–30 times higher than the dissolution fluid RB for the RB-labeled LC-MIP and LC-free MIP. The floating LC-MIP can remain in the stomach over 60 min, small intestine (60–90 min) and large intestine (180 min), respectively. On the other hand, the non-floating MIP remained in the stomach less than 30 min.

The LC-MIP showed a greater T_{max} (11.2 h) than the LC-free MIP (3.0 h), LC-NIP (1.2 h) and the commercial tablet, *S*-AML (2.5 h) (Fig. 5.6). For the LC-MIP, a plateau region of plasma concentration curve between 1.5 and 22 h indicated a equilibrium of the absorption and elimination. Due to the gastro-floating, a significantly higher of the value of $AUC_{0-\infty}$ (4180.4 and 3204.3 h ng/mL, respectively) for both LC-MIP and LC-NIP than LC-free MIP and the commercial tablet was observed, which had an $AUC_{0-\infty}$ of 2586.0, and 2316.1 h ng/mL, respectively. The relative bioavailability of the gastro-floating LC-MIP and

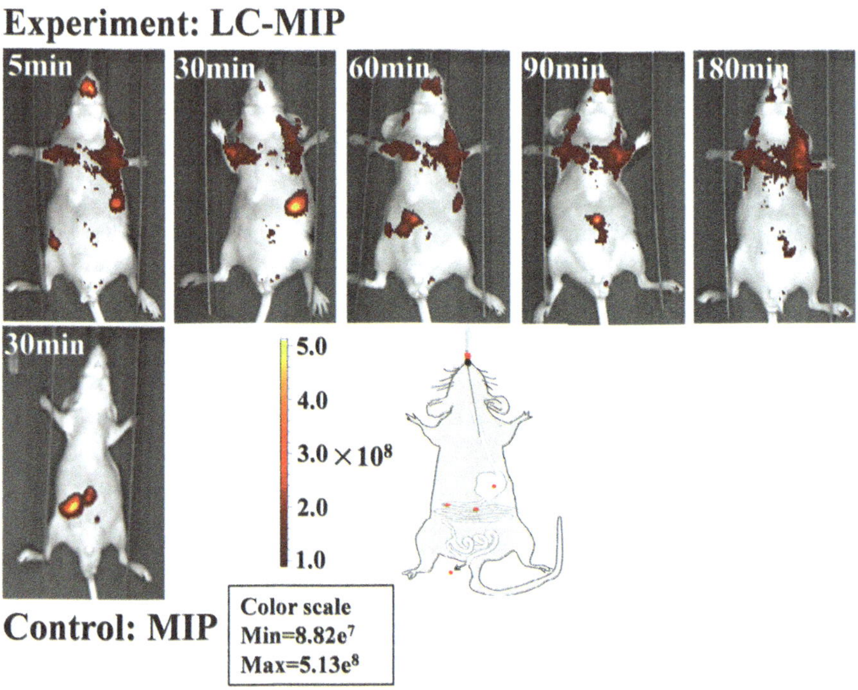

Fig. 5.5 Representative fluorescence images from nude mice following intragastrical administration of RB-labeled LC-MIP or LC-free MIP in void configuration. From left to right: Experimental group: 5, 30, 60, 90, 180 min after the floating LC-MIP (M3) intake; Control group: 30 min after the non-floating MIPs (M10) intake. (Reproduced with permission of [29])

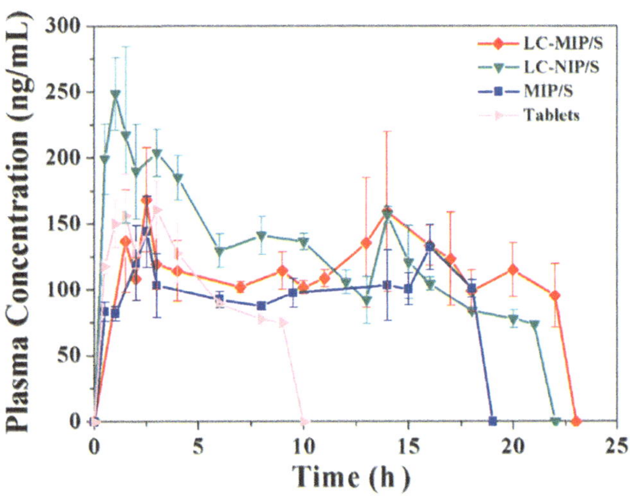

Fig. 5.6 In vivo pharmacokinetic profiles of *S*-AML for mice from LC-MIP, LC-NIP, LC-free MIP and the commercial tablet of *S*-AML. Data were presented as average, with error bars indicating the SD ($n = 3$ per group). (Reproduced with permission of [29])

LC-NIP was 180.5% and 138.3% compared with the conventional the commercial tablet, whereas LC-free MIP exhibited a relative bioavailability of only 111.7%. Therefore, the innovative combination of floating and controlled release properties indicated potentials of LC-MIPs for oral administration.

5.3.2 *Liquid Crystalline/Polyhedral Oligomeric Silsesquioxane Molecularly Imprinted Polymers for Gastroretentive Controlled Drug Release System*

Polyhedral oligomeric silsesquioxanes (POSS) are cagelike architecture-based organic-inorganic hybrids [31], and have attracted wide attention to construct multifunctional materials. As a porous inorganic core, the POSS monomer can be incorporated into polymers easily and result in noticeable improvement in many properties. Recently, it was found that nonspecific binding was suppressed when the POSS was incorporated into the matrix of MIPs [31–33]. Furthermore, cytocompatibility and nontoxicity of POSS monomers make them suitable for biomedical application.

A synergistic effect of POSS and LC was used to prepare MIPs [32] using methacryllsobutyl POSS, 4-methyl phenyl dicyclohexyl ethylene (MPDE) (LC monomer), methacrylic acid, and EDMA (Fig. 5.7). The simultaneous use of POSS and LC monomer also produced the changes in pore structure of the resulting MIPs different from the use of POSS or LC monomer independently.

The floating of the sustained drug delivery devices studied here was evaluated by investigating the gastroretentive effect of different polymers (Fig. 5.8) [33]. Here, the drug absorption or gastric emptying can be assessed by the decrease or disappearance of the fluorescence signal. The strongest fluorescence intensity of the drug from the conventional MIPs in the rat stomachs was observed after 0.5 h of administration. After 1.5 h, the very weak fluorescence intensity from the conventional MIPs was observed, and almost no fluorescent substance was detected by 2 h. In contrast, the LC POSS MIPs retained in the stomachs longer than the LC-MIPs, POSS MIPs as well as traditional MIPs. The signal intensity of the LC POSS MIPs disappeared until 2.5 h, indicating that the LC POSS MIPs had an enhanced effect of floating, which could improve the ability of drug release of the LC POSS MIPs materials and achieve better sustained release.

Paclitaxel (PTX) is a diterpenoid natural product of broad spectrum antitumor with terminal half-life of 1.3–8.6 h. More than 90% of the drug binds to plasma proteins rapidly and extensively, but approximately 50% onto red blood cells. However, to prepare DDS that can be taken orally it is very necessary because of its low aqueous solubility.

As shown in the profiles of plasma concentration versus time, the LC POSS MIPs and LC POSS NIPs gave the longest T_{max} of 3.0 h. The C_{max} of PTX for the LC POSS MIPs was slightly higher (50.0 ng/mL) than that of the LC POSS-free MIPs,

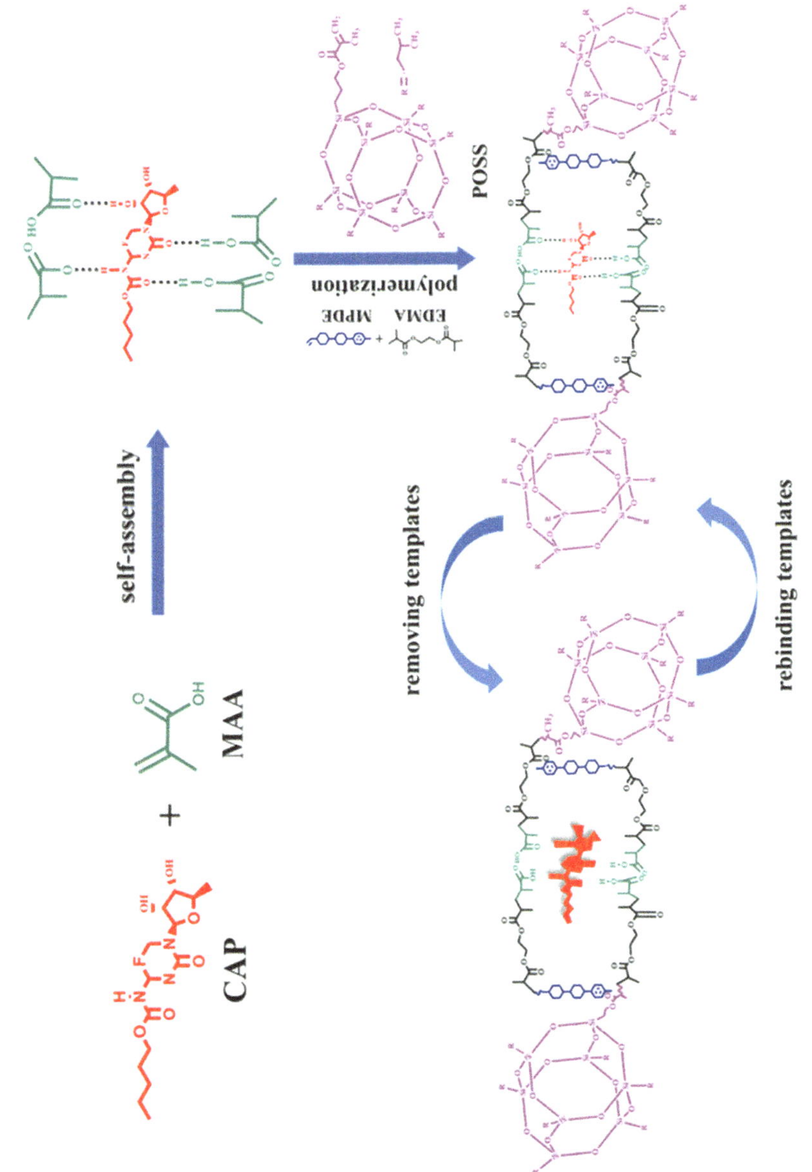

Fig. 5.7 Schematic representation of LC POSS MIPs preparation. (Reproduced with permission of [33])

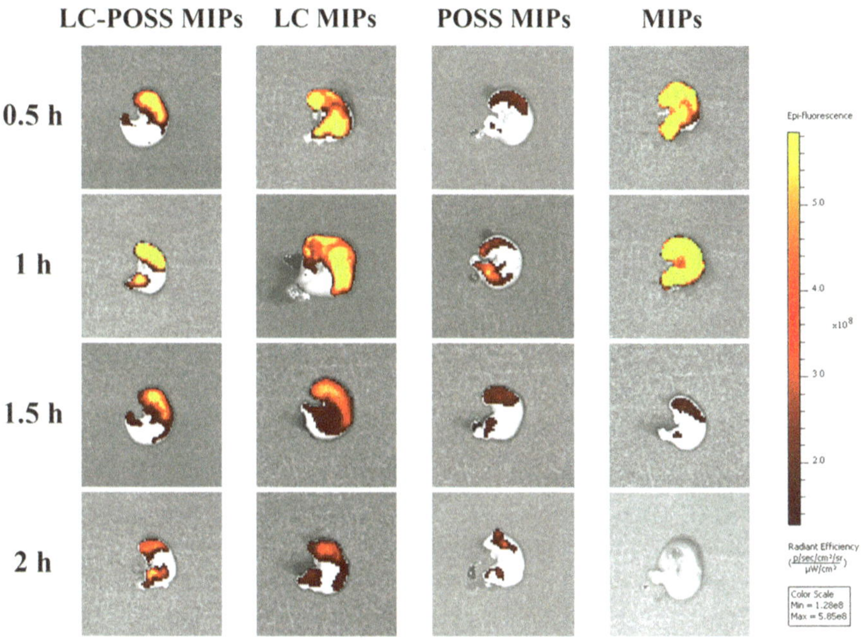

Fig. 5.8 Fluorescent images of the polymers in the stomachs of rats ($n = 3$). (Reproduced with permission of [33])

POSS LC-free MIPs and the pure PTX (41.6, 30.7 and 47.5 ng/mL), but slightly lower than the conventional MIPs (59.8 ng/mL). However, the highest relative bioavailability of the LC POSS MIPs (242.5%) was achieved, while the conventional MIPs, LC POSS NIPs, LC POSS-free or POSS LC-free MIPs was 84.6%, 152.5%, 127.4% and 63.2%, respectively. Furthermore, the LC POSS MIPs, LC POSS NIPs, LC POSS-free MIPs, POSS LC-free MIPs and conventional MIPs did not present any cytotoxicity to human breast cancer cell (MCF-7 cell). The cytotoxicity of each polymer was even lower than the drug-loaded polymer.

5.3.3 Liquid Crystalline/Metal Organic Framework Molecularly Imprinted Polymers for Gastroretentive Controlled Drug Release System

Metal organic frameworks (MOFs) are a recently identified class of porous polymeric materials, consisting of metal ions linked together by organic bridging ligands [34]. Flexibility of changing the metal centers and organic ligands enables the MOFs to be produced with multiple molecular functionalities and architectures. Their extraordinarily high surface areas, tunable pore sizes, ultrahigh porosities, and modifiable internal surfaces make them particularly attractive for catalyst, separation, gas storage, enrichment and sensing protein, as well as drug delivery.

Compared with other drug carriers, MOFs can bind drugs by anions–cations electrostatic interaction or hydrogen bonds, thus resulting in stable drug release effects. Furthermore, the low cytotoxicity and biodegradability of MOFs are favorable in the application of DDS.

$[Cu_3(BTC)_2(H_2O)_3]_n$, known as HKUST-1, is one of the reported MOFs that can be a good choice as carrier for drug delivery because of its unique large pore volume, very high surface area, good diffusion, and excellent antibacterial activity. However, the low permittivity and poor stability of neat HKUST-1 in water limited the use in DDS. To address these drawbacks, a novel MIP coated HKUST-1 containing a LC monomer was made (HKUST-1@LC-MIP). LC monomer, MPDE was also used to increase the floating of the composite. For example, the HKUST-1@LC-MIP and HKUST-1@LC-free MIP can float on the solution of pH 7.0 or 1.0 at least 24 h.

By contrast, the HKUST-1 sank immediately and the changed color of the solid in solution also suggested the varied surface morphology. Obviously, the stability of the HKUST-1@LC-MIP composite in water was improved by the existence of the MIP layers (Fig. 5.9). While soaked in the pH 1.0 solution long enough, however,

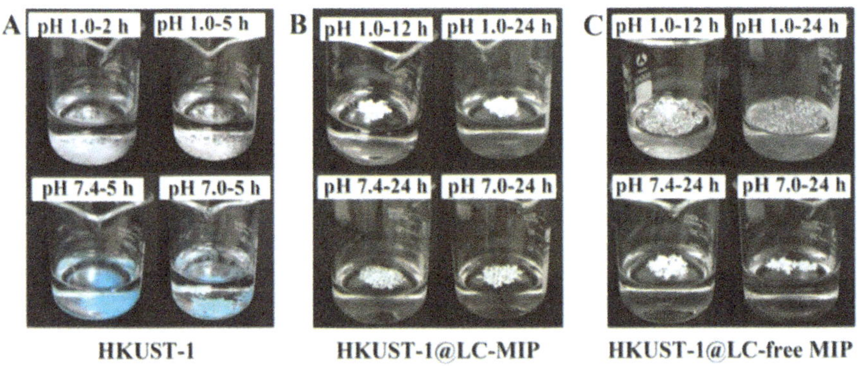

Fig. 5.9 The morphological changes of (**a**) HKUST-1, (**b**) HKUST-1@LC-MIP, and (**c**) HKUST-1@LC-free MIP after being soaked with simulated gastric juice (pH 1.0), intestinal fluid (pH 7.4), and the saline (pH 7.0). (Reproduced with permission of [35])

the bond of Cu–O of the HKUST-1 was broken, leading to the structure collapse of HKUST-1.

At the loaded concentration of 500 μg mL^{-1}, the HKUST-1@LC-MIP showed zero-order release of capecitabine (CAPE). In contrast, release of CAPE from the HKUST-1@LC-free MIP and HKUST-1@LC-free NIP completed 90% within 5 h. For the commercial CAPE tablets, the release percent reached 100% in about 7 h.

The HKUST-1@LC-MIP gave rise to higher relative bioavailability. As shown in the plasma concentration–time curves, the CAPE concentration in plasma reached a plateau between 8.0 and 12.0 h, indicating the adsorption/elimination equilibrium between CAPE and the HKUST-1@LC-MIP due to the controlled release during this period. For the HKUST-1@LC-NIP, the CAPE concentration declined quickly after administration. Specifically, the relative bioavailability (F%) from the HKUST-1@LC-MIP was 99.5% while that from HKUST-1 was only 50.3%.

5.3.4 Liquid Crystalline/Carbon Nanotubes Molecularly Imprinted Polymers for Gastroretentive Controlled Drug Release System

Multiwalled carbon nanotubes (MWCNTs) are attractive in DDS due to large surface areas, high strength, lack of swelling as well as stability under acidic conditions [36]. Nevertheless, the cytotoxicity of pristine CNTs would cause inflammation to human organs [37]. In contrary, the less toxicity of functionalized CNTs was observed [38], which can be applied to the targeted delivery. It was also found that great drug release performance of the functionalized CNTs DDS can be achieved compared with free drugs.

In general, MWCNTs-based MIPs are achieved by surface imprinting method. CNTs are used as support material and "grafting to" or "grafting from" approach is adopted to prepare MIPs. "Grafting to" approach involved covalent linkages of polymer chains or oligomer with the double bonds on the surface of CNTs. "Grafting to" approach, including noncovalent modification using surfactants, aromatic molecules, polymers and biomolecules, was adopted to make vinyl group functionalized MWCNTs with high density. Very thin MIP coating (ca. 15–20 nm) grafted to the surface of MWCNTs can be obtained, which led to less time of adsorption equilibrium to be achieved (within 35 min) because of the more accessible recognition sites to the template molecules. Thus, excellent binding kinetics can be achieved for MWCNTs@MIPs. A number of applications of MWCNT@MIPs including solid-phase extraction, sensors and drug delivery devices have also been found. However, only monomers containing reactive groups may be used for the technology.

Fabricating vinyl group functionalized MWCNTs was found proper for grafting LC-MIPs on the surface of MWCNTs (Fig. 5.10) [39]. With addition of 9-vinylanthracene, a high-density vinyl group functionalized MWCNTs was firstly obtained, and then polymerization for LC-MIPs was performed on the surface of

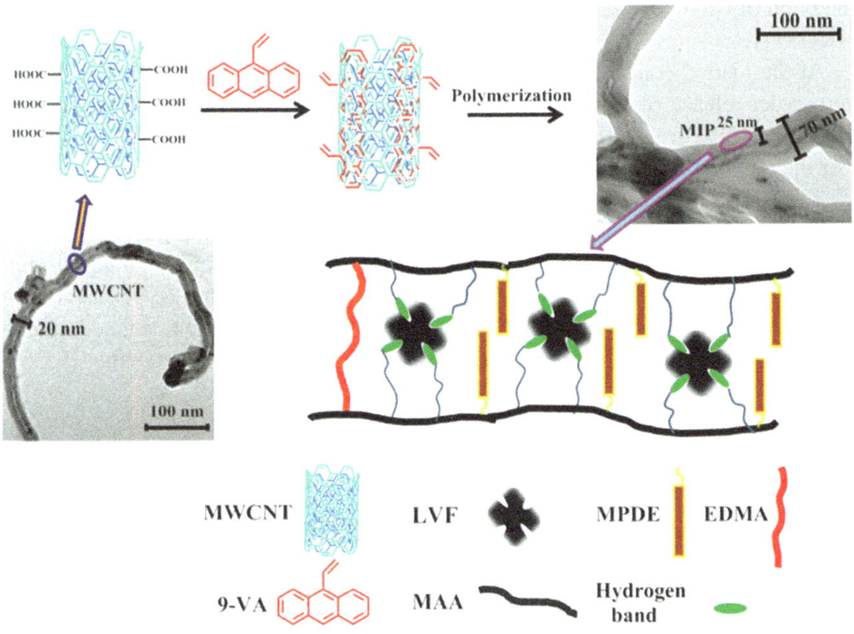

Fig. 5.10 Schematic representation of MWCNT@LC-MIP preparation. (Reproduced with permission of [39])

MWCNTs (MWCNT@LC-MIP). When introduced into the aqueous medium, the MWCNT@LC-MIP floated directly to upper surface, remaining buoyant for at least 24 h. The amounts of levofloxacin (LVF) adsorbed on the MWCNT@LC-MIP were 5.5-fold increase than that on the MWCNT@MIP.

For the MWCNT@LC-MIP, the adsorption rate was rapid at the first 60 min because of higher accessibility to the sites for LVF. The adsorption equilibrium was achieved after 30 min, suggesting easy access of LVF to the MWCNT@LC-MIP at the initial stage. In contrast, a longer time to the equilibrium (ca. 120 min) was required for the MWCNT@LC-free MIP.

In vitro release studies showed an obvious zero-order release of LVF from the MWCNT@LC-MIP with duration of about 20 h, at 3.8 μg/h of the release rate. For the MWCNT@LC-MIP, a relative plateau region of the LVF concentration in plasma between 1.0 and 10 h was displayed above the C_{max} of the commercial tablet. Enhanced bioavailability for the MWCNT@LC-MIP and MWCNT@LC-NIP was observed with a significantly higher value of $AUC_{0–13}$ (928.6 and 519.6 ng mL^{-1} h, respectively), than the commercial sustained tablets, MWCNT@MIP, and MWCNT and with $AUC_{0–13}$ value of 93.1, 18.7 and 160.4 ng mL^{-1} h, respectively. The relative bioavailability of the gastro-floating MWCNT@LC-MIP was 578.9%, whereas only 58.0% of MWCNT@MIP and 11.7% of the bared MWCNT.

5.4 Conclusion

It is clear that molecular imprinting provides a new strategy for stimuli responsive DDS. The MIP can act as thermo-responsive or pH-sensitive DDS with a longer period of time released. Nanomaterial@LC-MIP also was the innovative combination of floating and controlled release properties. As a whole, the MIP-based stimuli responsive DDS are very promising combination devices possessing the properties of controlled release and potentials for oral administration.

References

1. Pauling LJ (1940) A theory of the structure and process of formation of antibodies. J Am Chem Soc 62:2643–2657
2. Wei ZH, Sun X, Mu LN, Huang YP, Liu ZS (2019) Improving affinity of imprinted monolithic polymer prepared in deep eutectic solvent by metallic pivot. J Chromatogr A 1602:48–55
3. Song WF, Zhao QL, Zhou XJ, Zhang LS, Huang YP, Liu ZS (2018) A star-shaped molecularly imprinted imprinted polymer derived from polyhedral oligomeric silsesquioxanes with improved site accessibility and capacity for enantiomeric separation via capillary electrochromatography. Microchim Acta 186:22
4. Wu C, He J, Li Y, Chen N, Huang Z, You L, He L, Zhang S (2018) Solid-phase extraction of aflatoxins using a nanosorbent consisting of a magnetized nanoporous carbon core coated with a molecularly imprinted polymer. Microchim Acta 185:515
5. Yuan Y, Yang YJ, Zhu GS (2020) Molecularly imprinted porous aromatic frameworks for molecular recognition. ACS Cent Sci 6:1082–1094
6. Chen JF, Garcia ES, Zimmerman SC (2020) Intramolecularly cross-linked polymers: from structure to function with applications as artificial antibodies and artificial enzymes. Acc Chem Res 53:1244–1256
7. Liu GY, Huang XD, Li LY, Xu XM, Zhang YG, Lv J, Xu DH (2019) Recent advances and perspectives of molecularly imprinted polymer-based fluorescent sensors in food and environment analysis. Nanomaterials 9:1030
8. Wang X, Yang FF, Zhang LP, Huang YP, Liu ZS (2018) A polyhedral oligomeric silsesquioxane/molecular sieve codoped molecularly imprinted polymer for gastroretentive drug-controlled release in vivo. Biomater Sci 6:3170–3177
9. Tang L, Zhao CY, Wang XH, Li RS, Yang JR, Huang YP, Liu ZS (2015) Macromolecular crowding of molecular imprinting: facile pathway to produce drug delivery devices for zero-order sustained release. Int J Pharm 496:822–833
10. Liu XL, Yao HF, Chai MH, He W, Huang YP, Liu ZS (2018) Green synthesis of carbon nanotubes-reinforced molecularly imprinted polymer composites for drug delivery of fenbufen. AAPS PharmSciTech 19:3895–3906
11. Zaidi SA (2016) Molecular imprinted polymers as drug delivery vehicles. Drug Deliv 23:2262–2271
12. Rostamizadeha K, Vahedpourb M, Bozorgib S (2012) Synthesis, characterization and evaluation of computationally designed nanoparticles of molecular imprinted polymers as drug delivery systems. Int J Pharm 424:67–75
13. Balaure PC, Grumezescu AM (2015) Smart synthetic polymer nanocarriers for controlled and site-specific drug delivery. Curr Top Med Chem 15:1424–1490
14. Xu S, Lu H, Zheng X, Chen L (2013) Stimuli-responsive molecularly imprinted polymers: versatile functional materials. J Mater Chem C 1:4406–4422

15. Parisi OI, Ruffo M, Malivindi R, Vattimo AF, Pezzi V, Puoci F (2020) Molecularly imprinted polymers (MIPs) as theranostic systems for sunitinib controlled release and self-monitoring in cancer therapy. Pharmaceutics 12:41
16. Hemmati K, Sahraei R, Ghaemy M (2016) Synthesis and characterization of a novel magnetic molecularly imprinted polymer with incorporated graphene oxide for drug delivery. Polymer 101:257–268
17. Mao C, Xie X, Liu X, Cui Z, Yang X, Yeung KWK, Pan H, Chu PK, Wu S (2017) The controlled drug release by pH-sensitive molecularly imprinted nanospheres for enhanced antibacterial activity. Mater Sci Eng C Mater Biol Appl 77:84–91
18. Marcelo G, Ferreira IC, Viveiros R, Casimiro T (2018) Development of itaconic acid-based molecular imprinted polymers using supercritical fluid technology for pH-triggered drug delivery. Int J Pharm 542:125–131
19. Tashakori-Sabzevar F, Mohajeri SA (2015) Development of ocular drug delivery systems using molecularly imprinted soft contact lenses. Drug Dev Ind Pharm 41:703–713
20. Bordat A, Boissenot T, Nicolas J, Tsapis N (2019) Thermo responsive polymer nanocarriers for biomedical applications. Adv Drug Deliv Rev 138:167–192
21. Nagase K, Yamato M, Kanazawa H, Okano T (2018) Poly(N-isopropylacrylamide)-based thermoresponsive surfaces provide new types of biomedical applications. Biomaterials 153:27–48
22. Truong NP, Whittaker MR, Anastasaki A, Haddleton DM, Quinn JF, Davis TP (2016) Facile production of nanoaggregates with tuneable morphologies from thermoresponsive p(DEGMA-co-HPMA). Polym Chem 7:430–440
23. Ayari MG, Kadhirvel P, Favetta P, Plano B, Dejous C, Carbonnier B, Agrofoglio LA (2019) Synthesis of imprinted hydrogel microbeads by inverse Pickering emulsion to controlled release of adenosine 5′-monophosphate. Mater Sci Eng C Mater Biol Appl 101:254–263
24. Wang XL, Yao HF, Li XY, Wang X, Huang YP, Liu ZS (2016) pH/temperature-sensitive hydrogel-based molecularly imprinted polymers(hydroMIPs) for drug delivery by frontal polymerization. RSC Adv 6:94038–94047
25. Adali-Kaya Z, Tse SBB, Falcimaigne-Cordin A, Haupt K (2015) Molecularly imprinted polymer nanomaterials and nanocomposites: atom-transfer radical polymerization with acidic monomers. Angew Chem Int Ed Engl 54:5192–5195
26. Nuvoli D, Alzari V, Pojman JA, Sanna V, Ruiu A, Sanna D, Malucelli G, Mariani A (2015) Synthesis and characterization of functionally gradient materials obtained by frontal polymerization. ACS Appl Mater Interfaces 7:3600–3606
27. Wang C, Javadi A, Ghaffari M, Gong S (2010) A pH-sensitive molecularly imprinted nanospheres/hydrogel composite as a coating for implantable biosensors. Biomaterials 31:4944–4951
28. Wei P, Song R, Chen C, Li Z, Zhu Z, Li S (2019) A pH-responsive molecularly imprinted hydrogel for dexamethasone release. J Inorg Organomet Polym Mater 29:659–666
29. Zhang LP, Wang XL, Pang QQ, Huang YP, Tang L, Chen M, Liu ZS (2017) Solvent-responsive floating liquid crystalline-molecularly imprinted polymers for gastroretentive controlled drug release system. Int J Pharm 532:365–373
30. Ohm C, Brehmer M, Zentel R (2010) Liquid crystalline elastomers as actuators and sensors. Adv Mater 22:3366–3387
31. Li F, Chen XX, Huang YP, Liu ZS (2015) Preparation of polyhedral oligomeric silsesquioxane based imprinted monolith. J Chromatogr A 1425:180–188
32. Zhang LP, Tang SH, Mo CE, Wang C, Huang YP, Liu ZS (2018) Synergistic effect of liquid crystal and polyhedral oligomeric silsesquioxane to prepare molecularly imprinted polymer for paclitaxel delivery. Eur Polym J 98:226–236
33. Mo CE, Chai MH, Zhang LP, Ran RX, Huang YP, Liu ZS (2019) Floating molecularly imprinted polymers based on liquid crystalline and polyhedral oligomeric silsesquioxanes for capecitabine sustained release. Int J Pharm 557:293–303

34. D'Vries RF, Iglesias M, Snejko N, Gutieírrez-Puebla E, Monge MA (2012) Lanthanide metal-organic frameworks: searching for efficient solvent-free catalysts. Inorg Chem 51:11349–11355
35. Zhang LP, Mo CE, Huang YP, Liu ZS (2018) Preparation of liquid crystalline molecularly imprinted polymer coated metal organic framework for capecitabine delivery. Part Part Syst Charact 36:1800355
36. Wong B, Yoong SL, Jagusiak A, Panczyk T, Ho HK, Ang WH, Pastorin G (2013) Carbon nanotubes for delivery of small molecule drugs. Adv Drug Deliv Rev 65:1964–2015
37. Lanone S, Andujar P, Kermanizadeh A, Boczkowski J (2013) Determinants of carbon nanotube toxicity. Adv Drug Deliv Rev 65:2063–2069
38. Hwang JY, Nish A, Doig J, Douven S, Chen CW, Chen LC, Nicholas RJ (2008) Polymer structure and solvent effects on the selective dispersion of single-walled carbon nanotubes. J Am Chem Soc 130:3543–3553
39. Zhang LP, Tan XX, Huang YP, Liu ZS (2018) Floating liquid crystalline molecularly imprinted polymer coated carbon nanotubes for levofloxacin delivery. Eur J Pharm Biopharm 127:150–158

Chapter 6
MIP as Drug Delivery Systems for Dermal Delivery

Zehui Wei, Lina Mu, and Zhaosheng Liu

6.1 General Aspects of Transdermal Delivery System

Transdermal delivery system (TDS) or transdermal absorption preparation refers to the use of pharmacological, physical, and chemical methods to promote drugs to enter the circulatory system at a constant rate (or close to a constant rate) through the skin surface to produce systemic or local therapeutic effect. TDS in a broad sense includes patches, ointments, plasters, liniments, aerosols, etc. Some shortcomings caused by oral or subcutaneous injection, such as the first-pass effect of the liver, the pain and the risk of disease transmission due to the use of needle can be avoided in the transdermal delivery system. In addition, transdermal systems have the advantages of noninvasive, well patient compliance, long-acting, low-frequency administration, and cost-effective [1]. These excellent properties make transdermal delivery receive continuous attention and research as an important contribution to medical practice [2].

Transdermal drug products are usually adhered to the unbroken skin surface to deliver the active ingredient(s) to the stratum corneum, epidermis, dermis, muscle, and/or joint for the local therapeutic, or passing through the skin barrier for the systemic therapeutic. Skin structure is showed in Fig. 6.1. In recent years, the development of many new transdermal products has focused on promoting the penetration of drugs through the skin barrier, including small, lipophilic, low-dose

Z. Wei
School of Pharmacy, Jiangsu Ocean University, Jiangsu, China

School of Pharmacy, Tianjin Medicine University, Tianjin, China

L. Mu
Lianyungang TCM Branch of Jiangsu Union Technical Institute, Jiangsu, China

Z. Liu (✉)
School of Pharmacy, Tianjin Medicine University, Tianjin, China
e-mail: liuzhaosheng@tmu.edu.cn

Z. Liu et al. (eds.), *Molecularly Imprinted Polymers as Advanced Drug Delivery Systems*, https://doi.org/10.1007/978-981-16-0227-6_6

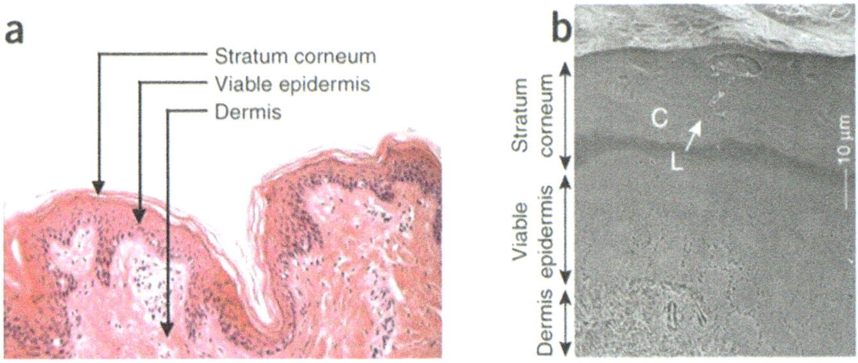

Fig. 6.1 (**a**) Skin structure: includes stratum corneum (acting as a skin barrier function), viable epidermis (renewing the stratum corneum) and dermis (providing skin's mechanical support, vasculature and anchoring of sweat gland and hair follicle appendages). (**b**) Stratum corneum structure. The lipid bilayer structure (L) plays a major role in restricting drug penetration. It exists in the intercellular space between corneocytes (C). (Reproduced with permission of [1])

drugs, even macromolecules and vaccines. These means used involve chemical enhancers, (non)-cavitational ultrasound, iontophoresis, microneedles, thermal ablation, microdermabrasion, electroporation, etc. [1]. However, in the development of transdermal products, the drug loading is also an indispensable factor, because this directly affects the longer duration of action. MIP is appealing as drug delivery vehicles, because MIP can inherently act as reservoirs via its high adsorption capacity and delay drug release via the special bonding force between template molecules and polymer materials [3, 4]. This provides new carrier materials for the development of long-acting and controlled/sustained-release transdermal products.

So far, the MIP-based transdermal delivery system that has been successfully developed includes but is not limited to the following forms, analog patch, composite imprinted membrane, mungbean starch/biomaterial imprinted films, and gel reservoir-type transdermal patch based on imprinted membrane.

6.2 Analog Patch

Transdermal patches usually contain an outer covering, which mainly supports active substance(s) and several kinds of excipients. Common excipients include stabilizers, solubilizes, release rate regulators, transdermal absorption enhancers, etc. According to the design, the transdermal patches are usually divided into three types: reservoir, matrix, and drug-in-adhesive (Fig. 6.2). Currently, a large number of transdermal patches have been successfully developed and commercialized, such as lidocaine, nicotine, buprenorphine, fentanyl, estradiol, granisetron, testosterone, levonorgestrel, nitroglycerin, methylphenidate, oxybutynin, scopolamine, clonidine,

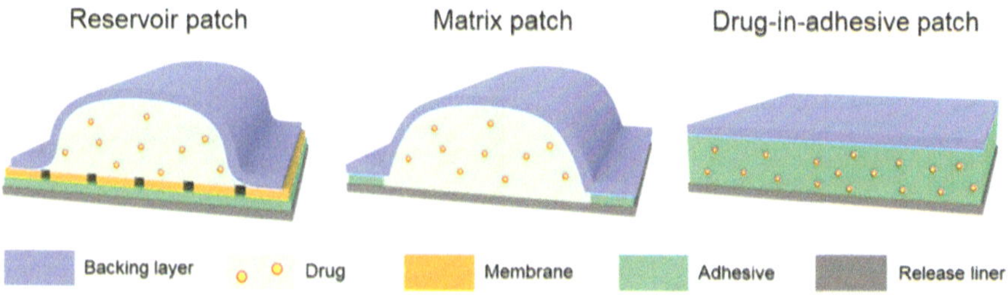

Fig. 6.2 Schematic illustration of reservoir patch, matric patch and drug-in-adhesive patch. (Reproduced with permission of [5])

and so on [5]. The advantages of painless, comfortable, low first-pass effect, and side effects make transdermal patches widely used in clinical treatment.

However, in transdermal patches, the active substances tend to penetrate the polymer matrix easily, which leads to difficult control their release rate from the device [6]. Coincidentally, the imprinted polymer materials with special release properties can be a powerful solution to control the release rate of active substances.

Nicotine transdermal delivery has been proved to be an effective therapy for smoking cessation, and nicotine injection or skin patches display significant effects in improving the attention, learning, and memory of Parkinson's patients [7, 8]. Nicotine MIP can be applied as a safe and effective matrix to control the sustained release of drugs. The preparation and evaluation of MIP-based nicotine transdermal patche mainly accord to the research reports of Ruela et al. [9].

6.2.1 Nicotine Transdermal Patch Based on Bulk Polymerization

In the first step, the nicotine-imprinted copolymers were prepared through in situ bulk polymerization. The pre-polymerized solution containing nicotine (1.0 mmol), methacrylic acid (MAA, 4.0 mmol), ethylene glycol dimethacrylate (EDMA, 20 mmol), azobisisobutyronitrile (0.24 mmol) and methylene chloride (6 mL) was dissolved by ultrasound, degassed by purging nitrogen and sealed in a reaction flask. After overnight polymerization at 60 °C, the resulting imprinted material was crushed, ground and sieved (75–106 μm). Then, a certain amount of polymer particles were dispersed in 5% (w/w) drug solution in vehicles of mineral oil (NP, non-polar) or propylene glycol (P, polar) and coated onto aluminum foils disks with the size of 1.8 cm^2. The pH of nicotine analog patch was adjusted to about 6.0, which was similar to pH of the skin. The schematic diagram of nicotine imprinted polymers was shown in Fig. 6.3a.

The vertical Franz-type glass diffusion cell apparatus were employed to evaluate the transdermal delivery of nicotine analog patch. The porcine ear skin was fixed between the donor cell and receptor cell with the stratum corneum facing the donor

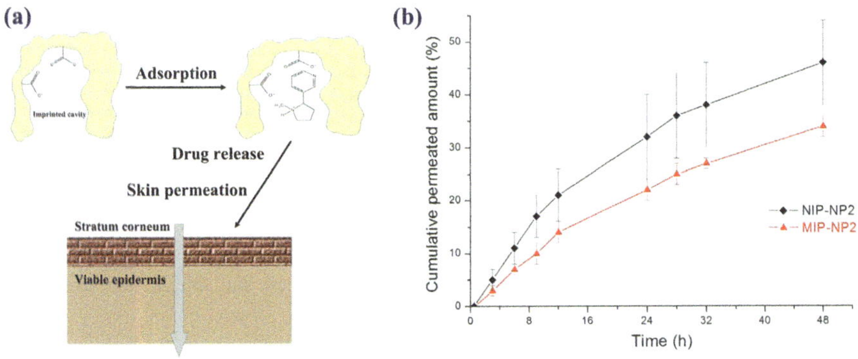

Fig. 6.3 (**a**) The schematic diagram of nicotine imprinting polymers and transdermal delivery; (**b**) the cumulative permeated amount of nicotine on MIP/NIP-based transdermal patch, NP means mineral oil as vehicles. (Reproduced with permission of [9])

compartment. The nicotine analog patch was directly covered onto the surface of skin in donor cell, and the isotonic PBS buffer (pH 7.4, 20 mM) was filled in receptor chamber with a 600 rpm stirring rate. The whole transdermal experiment was kept at 32 ± 1 °C. One milliliter aliquots were took out at 0.5, 3, 6, 9, 12, 24, 28, 32, 48 h time intervals and immediately equal volume of phosphate buffer was added into receptor chamber.

According to the results of transdermal experiments, the cumulative permeated amount of nicotine on MIP-based transdermal patch was significantly smaller than that of NIP-based transdermal patch. This also clearly showed that the diffusion of drugs from the polymer matrix was regulated by specific imprinting sites. The release data were better fitted to the Higuchi model, indicating that the diffusion from the polymer matrix was the main factor for nicotine release (Fig. 6.3b).

$$\text{Higuchi's model}: \frac{Q_t}{Q_0} = K_H \times t^{1/2} \tag{6.1}$$

Q_t represents the amount (μg) of diffusion at time t (h); Q_0 represents initial amount (μg) in donor cell (μg); K_H represents Higuchi's constant (μg $h^{1/2}$).

Two other commonly used kinetic models are as follows.

$$\text{Zero order model}: Q_t = Q_0 + K_0 t \tag{6.2}$$

$$\text{First order model}: LogQ_t = LogQ_0 + K_1 t \tag{6.3}$$

where K_0 represents zero order constant (μg h^{-1}), K_1 represents first order constant (μg h^{-1}).

For the cumulative amount of nicotine transdermal delivery at 24 h, MIP-based transdermal patch can finally reached 655 ± 29 μg cm^{-2}. This result was quite close to the release of commercial patch Nicopatch® (14 mg/24 h) [10], which had the cumulative transdermal delivery of nicotine of 709 ± 56 μg cm^{-2} at 24 h. The above results indicate that the MIP can be implemented as a potential nicotine transdermal delivery system and functional pharmaceutical excipient to improve stability and control drug release.

6.2.2 *Nicotine Transdermal Patch Based on Precipitation Polymerization*

Ruela's group further prepared nicotine MIP by precipitation polymerization and assembled them into nicotine transdermal patches [11]. The preparation of nicotine MIP was as follows. MAA (7.5 mmol), 2-hydroxyethyl methacrylate (HEMA, 7.5 mmol), and nicotine (1.875 mmol) were dissolved in 150 mL toluene-acetonitrile mixed solvent (as porogenic solvent). After standing for 30 min, EDMA (30 mmol) and 1,1′-azobis(cyclohexanecarbonitrile) (ABCN, 0.2 mmol) were added to the above solution and degassed with N_2 for 15 min. The pre-polymerization solution was stirred overnight at 80 °C for reaction. The resultant polymers were successively rinsed with methanol: acetic acid (1:1, v/v) and water, and dried 24 h at 50 °C before use. The NIP was synthesized using the same preparation method as the MIP except for no adding nicotine in polymerization solution. Then, the nicotine transdermal patch was assembled. The MIP particles were dispersed in a certain volume of mineral oil, which contained a certain quality of nicotine. The dosage of the drugs and MIP particles were mainly selected based on the amount of nicotine released in the commercial patch, so as to release about 14 mg of drug in 24 h. Using aluminum foil (1.8 cm^2) as a backing layer, the mixture was coated onto disks for in vitro performance assays.

The skin, equipment, and evaluation methods used in the skin penetration test were consistent with the above test conditions. This patch penetrated nicotine in a zero order kinetic mode up to 48 h (Table 6.1), and the permeation flux for the MIP was significantly smaller than that of the NIP. This result showed that the MIP can control the release of nicotine and have more specific imprinting sites than the NIP. The permeation flux was also related to the amount of polymer. When the amount of the polymer used in the formulation increased, the permeability flux of the MIP and NIP decreased. This may be due to the swelling behavior of the polymer, which increased the barrier of the drug to the skin. Compared with the above bulk polymerization-based MIP, which followed Higuchi kinetic, the MIP produced by precipitation polymerization allowed the resultant transdermal patch with skin permeation following zero order kinetic. Under 30 mg polymer dosage, the drug release flux was 41.69 ± 1.16 μg cm^{-2} h^{-1} over 24 h from the matrix or 38.29 ± 3.56 μg cm^{-2} h^{-1} over 24 h from a membrane, which was very close to the release flux of commercial nicotine transdermal patches [12]. Therefore, when used as an advanced adjuvant to control the release of nicotine, the imprinted copolymers and their respective controls (NIP) can control the penetration of nicotine in the skin.

Table 6.1 The kinetic parameters of zero order model for in vitro skin penetration study of nicotine transdermal patch ($n = 3$). (Reproduced with permission of [11])

Formulation	Nicotine (mg)	Polymer (mg)	r	Flux ± SD ($\mu g\ cm^{-2}\ h^{-1}$)
MIP/mineral oil	10.0	50.0	0.99691 ± 0.00138	26.21 ± 3.33
NIP/mineral oil	10.0	50.0	0.99540 ± 0.00136	30.94 ± 2.40
MIP/mineral oil	10.0	30.0	0.99668 ± 0.00029	46.48 ± 0.37
NIP/mineral oil	10.0	30.0	0.99160 ± 0.00599	60.21 ± 3.12

In short, whether bulk polymerization or precipitation polymerization, the MIP obtained by these two methods can be applied as potential pharmaceutical excipients in the construction of controlled/sustained-release transdermal patches. The main function of MIP-based transdermal patch can be achieved as follows.

- Painless, comfortable, low first-pass metabolism and side effects;
- Avoid the burst effect caused by the excessive loaded of drugs when using other polymer carrier materials;
- Reduce drug loss during storage due to the drug volatility, e.g., nicotine;
- Control drug release, such as zero order kinetic, Higuchi kinetic;
- Improve compatibility with formulation components.

6.3 Composite Imprinted Membrane

The permeability and porous properties of the membrane provide a good carrier for drug delivery or release from the storage site [13]. The combination of MIP and membranes can provide new dosage forms for transdermal drug delivery. Several technologies have emerged to incorporate MIP into the membrane, such as coating or grafting MIP onto the surface of membrane disc [14, 15], pore-filling of thin track-etched membranes with MIP [16], encapsulation of MIP nanoparticles between two membrane layers or into a composite nanofiber membrane [17, 18].

Common polymers, such as polyamide [17], polysulfone [19], and cellulose acetate [16, 20], are employed as the membrane basis to prepare MIP composite membranes. However, these synthetic membranes are not suitable for transdermal delivery systems due to their biocompatibility issues and inappropriate physical properties. The bacterial-derived cellulose which was synthesized by Acetobacter xylinum has been used as a temporary skin substitute, surgical or microsurgery materials, dressings and membranes [21, 22]. This cellulose membrane is highly resistant to corrosive chemicals and environmentally friendly and biodegradable.

Therefore, the bacterial-derived cellulose membrane should be considered as an ideal composite imprinting membrane basis for transdermal delivery systems.

6.3.1 Surface Grafting-Based MIP Composite Membrane

Roongnapa Suedee's research group constructed an *S*-propranolol cellulose membrane with transdermal delivery function based on MIP and bacterial-derived membranes [23]. The surface grafting procedure of MIP onto the bacterially derived membrane was illustrated in Fig. 6.4a. Firstly, the cellulose membrane was derivatized with 3-methacryloxypropyltrimethoxysilane (3-MPS) (10% w/w in toluene) for 5 h at 80 °C. This made the membrane surface modification producing unsaturated double bonds for MIP by graft polymerization. The derivatized membrane was rinsed with methanol and dried, and placed in an 18 cm crystallizing dish. Then, the surface derivatized membrane was immersed in the pre-polymerization solution, which contained MAA (functional monomer), EDMA (cross-linkers), *R*- or *S*-propranolol (HCl) (template), AIBN (initiator), and DMF (solvent).

After reaction 18 h at 60 °C, the surface grafting-based *R*- or *S*-propranolol imprinted composite membrane was obtained under the sequential extraction with 10% acetic acid in methanol (v/v) and pure methanol using Soxhlet extractor. SEM cross-sectional images (Fig. 6.4b) could be clearly observed that the MIP layer was closely packed on the surface of the cellulose membrane.

R- or *S*-propranolol percutaneous permeation study in vitro via surface grafting-based MIP composite membrane was performed with a dorsal skin of Wistar rats by using Franz-type diffusion cell. The epidermal side was in contact with the MIP composite membrane, and the dermal surface was faced the receptor cell. Drug solutions with different concentrations were added into the donor cell, and an aliquot of receptor solution was collected at prescribed time intervals while the receptor phase was maintained at 37 °C and 250 rpm stirring. At the same time, an equal volume of phosphate buffer was replenished. The propranolol concentration in collected sample was determined by HPLC. For *S*-MIP membrane, NIP membrane and cellulose membrane, the transdermal delivery of *S*-propranolol was significantly greater than *R*-propranolol at all time points. This situation was consistent at 100, 200, 300 μg/mL drug loading doses. Moreover, the *S*-MIP membrane for improving the transdermal transmission of *S*-propranolol was significantly greater than the other two membranes (Fig. 6.5b–d). However, for the control group (using racemic propranolol solution directly), the transdermal delivery of *R*-propranolol was higher than that of *S*-propranolol (Fig. 6.5a). This result indicated that the cellulose membrane and MIP have a synergistic effect on improving the transdermal delivery of *S*-propranolol.

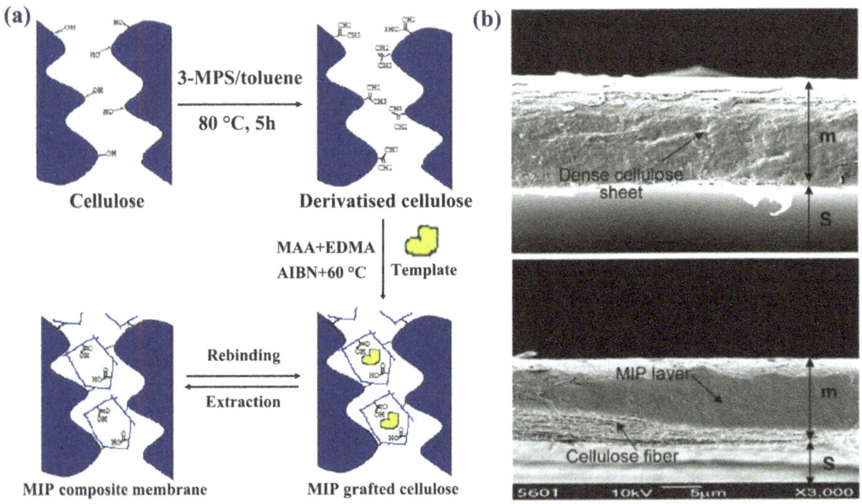

Fig. 6.4 The preparation procedure of surface grafting-based MIP composite membrane (**a**); the SEM cross-sectional image of cellulose membrane and MIP composite membrane (**b**). (Reproduced with permission of [23])

6.3.2 Nanoparticle-on-Microsphere-Based MIP Composite Membrane

MIP particles with heterogeneous morphology and extensive binding sites can be used as drug carriers to prepare imprinted composite membranes. However, the MIP particle-based composite membrane obtained by conventional methods has poor permeability and binding site accessibility, which is mainly caused by the close packing of particles. Jantarat et al. [24] employed a phase inversion technique to produce self-assembled MIP nanoparticle-on-microspheres (NOM)-based composite porous cellulose membranes. *S*-propranolol was used as a template molecule and polycapolactone-triol as plasticiser.

The preparation process included the following operations.

- **Preparation of MIP-NOM.** The pre-polymerization solution was prepared by dissolving 1.6 mmol MAA, 9.2 mmol EDMA, 0.4 mmol *S*-propranolol, 0.04 mmol AIBN, and 25 mg perfluoro polymeric surfactant (PFPS) in 20 mL perfluoro(methylcyclohexane) (PMC) and 5 mL chloroform. The mixed solution was degassed by N_2 stream for 5 min, and then vigorously stirred at 1000 rpm using a magnetic stirrer. The MIP-NOM was UV-initiated polymerization at 365 nm for 4 h. After polymerization, the MIP-NOM was washed three times using methanol-acetic acid (9:1, v/v) and methanol to remove template molecule, and then dried and stored at room temperature.
- **Casting of MIP Composite Membrane.** Firstly, 96 mg of bacterial cellulose sheet was dispersed in a 5 mL *N*-methylmorpholine-N-oxide (NMMO) solution

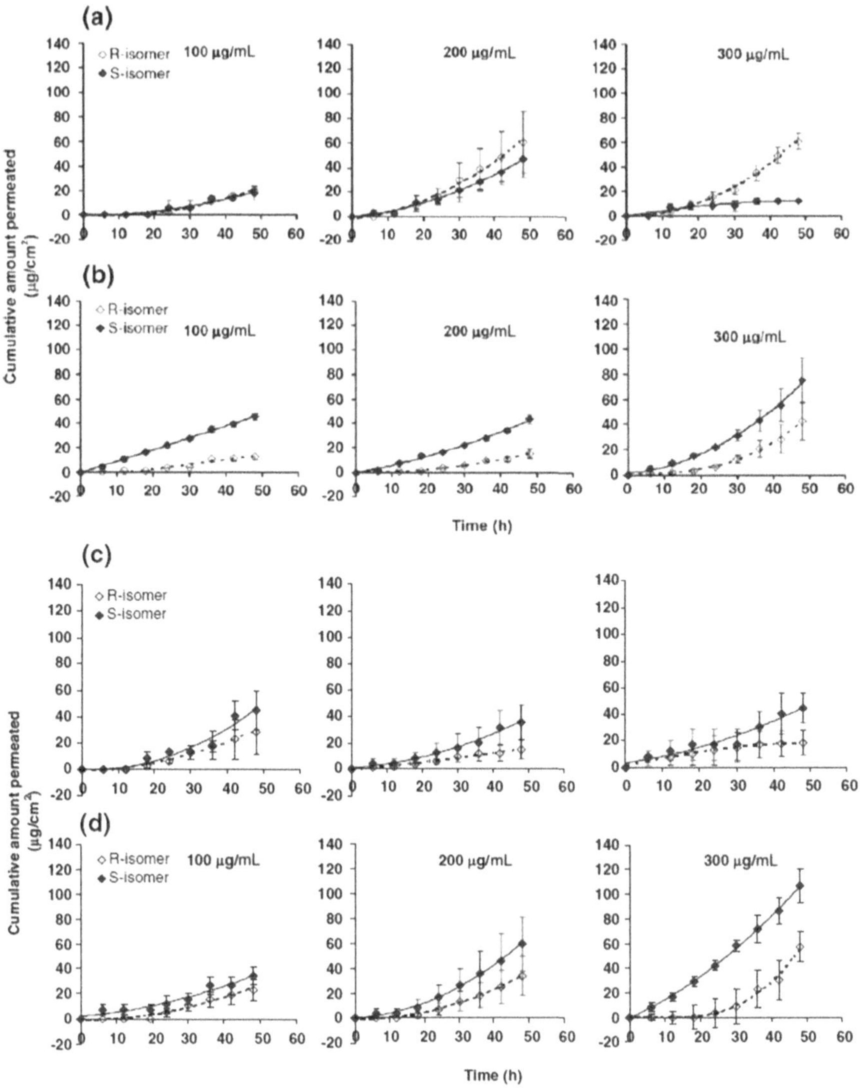

Fig. 6.5 Transdermal transfer curve of *S*-propranolol MIP cellulose membrane. (**a**) Control group (using racemic propranolol solution directly); (**b**) cellulose membrane; (**c**) NIP membrane; (**d**) *S*-MIP membrane. (Reproduced with permission of [23])

(50%, w/w) at 60 °C. This suspension was warmed to 80 °C until a clear, brown and viscous solution was formed. Then, 10 mg of racemic propranolol hydrochloride was mixed in sequence with 100 mg of the MIP-NOM and 0.3 mL of polycaprolactone-triol (PCL-T). Next, this mixture were added into the above cellulose solution. The obtained melted suspension was poured into a petri dish (diameter: 10 cm) for standing 5 min to cast into a membrane. The obtained membrane was transferred to 500 mL of distilled water and stood for 12 h. At last, the obtained membrane was cut into 28 cm^2 squares, dried in air, and stored in a desiccator.

SEM images of the MIP particles reflected that the stirring method had a crucial influence on the formation and morphology of the particles and was a key factor in the formation of the NOM-MIP. Rotating blade stirring could obtain irregular granules (Fig. 6.6a) and smooth microspheres (Fig. 6.6b). The magnetic bar stirring (4 cm $\times$ 1 cm) with a magnetic stirrer at 1000 rpm could form the desired NOM with rough surface (Fig. 6.6c). The surface (Fig. 6.6e) and cross-section (Fig. 6.6f) imaging of the NOM-MIP composite membrane clearly indicated that the NOM-MIP was evenly embedded in the cellulose membrane, which was significantly different from the imaging of the cellulose membrane alone.

Skin penetration test was measured by using male Wistar rat dorsal skin in a Franz-type diffusion cell. In the donor compartment, the composite membrane (0.8 cm^2) was fixed on the surface of the stratum corneum. The receiving compartment was filled with PBS buffer (pH 7.4, 2.5 mL), which was maintained at 37 $\pm$ 1 °C. 200 μL of samples from the receptor cell was collected within a predetermined time interval (0–48 h), and immediately an equal volume of PBS buffer was added. The enantiomeric release results indicated that the transdermal delivery of *S*-isomers drugs (including propranolol and other β-blockers, isovalerylpropranolol, cyclopropanoyl-propranolol, pindolol, and oxprenolol) from the NOM-MIP composite membranes was higher than that of *R*-isomer (Table 6.2). In contrast, the NOM-NIP composite membrane showed the same penetration rate for the *S*- and *R*-isomers of propranolol and other β-blockers. Therefore, the NOM-MIP-based composite membrane may be a potential transdermal preparation for stereoselective controlled delivery of propranolol and its prodrug analogs.

6.4 Mungbean Starch/Biomaterial Imprinted Films

In various renewable polymers, starch is a widely used and cheap natural biodegradable polymer because of the low cost of materials and simple processing equipment required to synthesize them [25, 26]. Starch-based functional biopolymer materials have been applied in wound dressings, drug carrier, and transdermal drug delivery patch [27, 28]. Recently, Tak [29] synthesized a biodegradable biomaterial imprinted films based on mungbean starch (MBS), polyvinyl alcohol (PVA), and plasticizers by UV irradiation. The nonsteroidal anti-inflammatory drugs (NSAIDs), sulindac (SLD), was used as template molecule, which has many side effects of stomach ulcers, digestive disorders, headache, itching, pancreatitis, sensitivity to light, cholestasis, hair loss, tinnitus, edema, high blood pressure, hematological disorders, palpitations, proteinuria, hematuria occurs and painful urination. The development of sulindac transdermal patch is conducive to control drugs release, reduce side effects and improve efficacy.

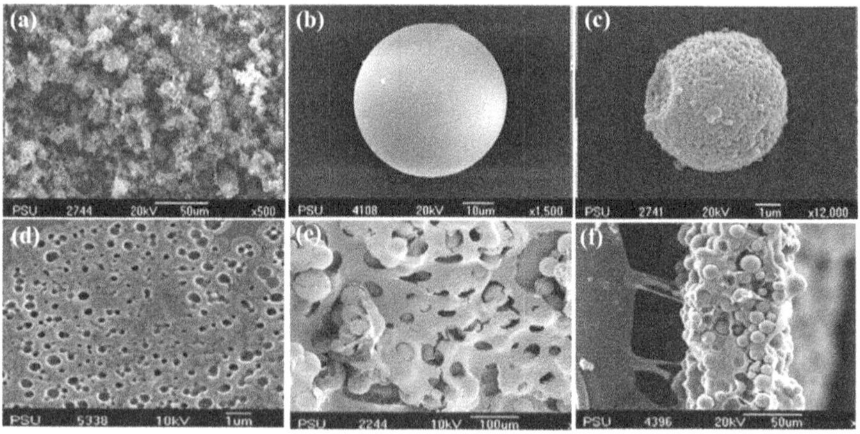

Fig. 6.6 SEM images of NOM-MIP composite membrane. Granules (**a**) and microspheres (**b**) prepared by rotating blade, nanoparticle-on-microspheres (**c**) prepared by magnetic bar (4 cm × 1 cm), blank cellulose cast membrane (**d**), surface (**e**), and cross-section (**f**) of NOM-MIP composite membrane. (Reproduced with permission of [24]

Table 6.2 In vitro rat skin permeation data of racemic propranolol, prodrugs of propranolol and other β-blockers release from MIP and NIP NOM based bacterial cellulose membranes in pH 7.4 phosphate buffer at 37 ± 1°C (mean ± S.D., $n = 3$). (Reproduced with permission of [24])

		J_{ss} (g cm^{-2} h^{-1})		
Drug	Membrane	*R*-isomer	*S*-isomer	*S*/*R* ratio
Propranolol	NIP	0.32 ± 0.06	0.31 ± 0.12	0.95 ± 0.18
	MIP	0.43 ± 0.09	0.57 ± 0.08	1.33 ± 0.18
Cyclopropanoyl-propranolol	NIP	2.17 ± 0.01	1.96 ± 0.03	0.91 ± 0.01
	MIP	8.72 ± 0.07	11.79 ± 0.13	1.36 ± 0.03
Valeryl-propranolol	NIP	4.86 ± 0.92	5.08 ± 0.33	1.05 ± 0.13
	MIP	4.62 ± 0.11	5.87 ± 0.02	1.27 ± 0.02
Oxprenolol	NIP	9.85 ± 3.63	10.44 ± 4.56	1.04 ± 0.14
	MIP	16.77 ± 0.88	19.48 ± 0.69	1.16 ± 0.02
Pindolol	NIP	5.04 ± 0.68	4.91 ± 0.74	0.97 ± 0.01
	MIP	3.83 ± 0.32	4.48 ± 0.26	1.16 ± 0.03

6.4.1 The Synthesis of MBS/PVA Biomaterials Imprinted Films

The synthesis of SLD imprinted MBS/PVA biomaterials films were divided into the following steps.

- Dissolved 5.0 g PVA in 90 °C hot water.
- Used a kitchen aid blender to mix 5.0 g MBS and 40% (wt%) plasticizer (glycerin-GL, citric acid-CA, or ascorbic acid-AsA) with water for 20 min.

- The MBS/plasticizers and PVA solution were heated to 95 °C respectively, and then stirred by mechanical stirring (600 rpm) for 60 min until a homogeneously gel-like solution formed.
- Dissolved 0.2 g SLD in 20 mL ethanol, and added it dropwise to the gel solution within 10 min. Removed bubbles using an aspirator.
- Then poured the gel-like solution to a preheated Teflon mold (60 °C, 250 × 250 × 1.0 mm). Moisture was evaporated by heating for 24 h in a 60 °C oven. The obtained biomaterial film was irradiated with ultraviolet light for 30 min under atmospheric pressure.
- After being irradiated with ultraviolet light, these imprinted films were humidity-conditioned again at RH 55% and 25 °C for 7 days.

6.4.2 The Properties Evaluation of MBS/PVA Biomaterials Imprinted Films

Common physical and chemical properties evaluation methods were as follows. Instron 6012 testing machine was employed to evaluate tensile strength (TS) and elongation at break (%*E*). The properties of solubility (*S*) and swelling behavior (SB) was tested by weight method. Firstly, the dried films were soaked in 25 °C distilled water for 24 h to achieve equilibrium. The weight of the film after swelling was then measured. The Eqs. (6.4) and (6.5) were used to calculate the SB and *S* of the each film.

$$\mathrm{SB} = \frac{W_\mathrm{e} - W_0}{W_0} \tag{6.4}$$

$$S = \frac{W_0 - W_\mathrm{d}}{W_\mathrm{d}} \tag{6.5}$$

Where W_0 represents the weight of the dry film; W_e represents the weight of the film when its adsorption reaches equilibrium; W_d represents the re-dried weight of the swelling balanced film which is dried for 24 h at 60 °C.

The Eq. (6.6) was used to calculate the gel fraction.

$$\text{Gel fraction } (\%) = \frac{W_\mathrm{g}}{W_0} \times 100 \tag{6.6}$$

where W_g represents the weight of the dry gel film after extraction. The extraction was done by placing MBS/PVA biomaterial imprinted films on a 200 mesh stainless steel net, which were immersed in DMSO solution for 72 h. After fully rinsed with distilled water and methanol, the films were dried at 50 °C to constant weight.

The MBS/PVA biomaterials imprinted films were prepared by ultraviolet light irradiation. The irradiation time had a great influence on the physical properties of the film. With increased irradiation time, TS gradually increased with rapid decrease of *%E* (Fig. 6.7a). When the irradiation time exceeded 30 min, the film was deformed, and became more brittle. This indicated that appropriate UV irradiation time (30 min) was a necessary condition for obtaining a film with good physical properties. Both SB and *S* decreased with the increase of UV exposure time (Fig. 6.7b). The increase in the degree of cross-linking was the main factor causing this phenomenon. Gel fraction is one of the important parameters to measure the cross-linking degree of hydrogel or films [30]. For the MBS/PVA imprinted films, the gel fraction increased sharply when the irradiation time reached about 30 min (Fig. 6.7c). These results indicated that UV irradiation indeed initiated the cross-linking reaction between the components. The surface of the pure MBS/PVA biomaterial film appeared relatively uniform and smooth, while the SLD imprinted film had a yellow appearance (Fig. 6.7d). All the MBS/PVA biomaterials imprinted films exhibited no obvious cracks, voids, peeling, or agglomeration.

6.4.3 The Release Properties of MBS/PVA Biomaterials Imprinted Films

For the release performance test, the MBS/PVA biomaterial imprinted film (0.10 g) was put into a volumetric flask containing a buffer solution of pH 4.0, 7.0, or 10.0. Then, it was fixed in a 37 °C shaking incubator (80 rpm) for incubation. 2.0 mL of the release solution was pipetted at predetermined time interval and it was analyzed at 327.3 nm by an ultraviolet-visible spectrophotometer, and immediately replenished an equal volume of the release medium. An artificial skin (NeodermR-ED, Tego Science, Inc. Korea) was used to evaluate the transdermal permeability of the patch at RH 60.0% and 36.5 °C. The results showed that the cumulative release of SLD under various conditions exceeded 95% within 10 h. The release effect of SLD in pH 7.0 and 10.0 buffer was better than pH 4.0, which was mainly due to the increase in the solubility of SLD with increasing pH (Fig. 6.8a) [31]. It can be seen from the release curve of SLD on human skin (pH 6.8 and 36.5 °C) that the imprinted membrane can quickly release SLD within 10 h, and the added plasticizer could slightly affect the release rate of SLD, with the order of MBSPCA4-SLD < MBSPAsA4-SLD < MBSPGL4-SLD < MBSP-SLD (Fig. 6.8b). The cumulative release rate of SLD increased with time at a relatively steady rate on artificial skin, and cumulative release over 24 days reached about 95.0–98.0% (Fig. 6.8c). It can be seen from the picture that after 20 days of continuous release, the SLD imprinted films remained light yellow, indicating that there was still some SLD waiting to be released. These results indicated that the developed MBS/PVA

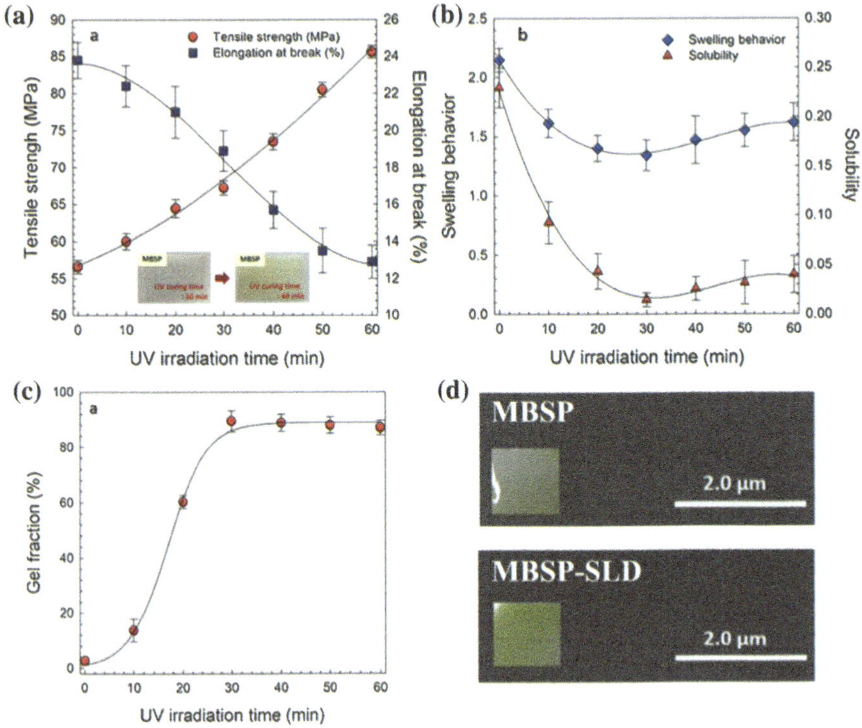

Fig. 6.7 The property evaluation of MBS/PVA biomaterials imprinted films. (**a**) Tensile strength (TS) and elongation at break (*%E*); (**b**) swelling behavior (SB) and solubility (*S*); (**c**) gel fraction; (**d**) SEM images. (Reproduced with permission of [29])

imprinted film could be a potential biomaterial to be applied as a long-term transdermal drug delivery patch.

6.5 Gel Reservoir-Type Transdermal Patch Based on Imprinted Membrane

Chitosan is a naturally occurring polysaccharide, mainly prepared by N-deacetylation of chitin [32]. The low toxicity and weak immunogenicity of chitosan make it a potential vehicle in transdermal delivery systems [33]. The Suedee's research group used chitosan as a drug reservoir and MIP thin-layer composite cellulose membrane as the enantioselective-controlled release material to construct a transdermal patch close to practical, and completed the in vivo evaluation of Wistar rats [34].

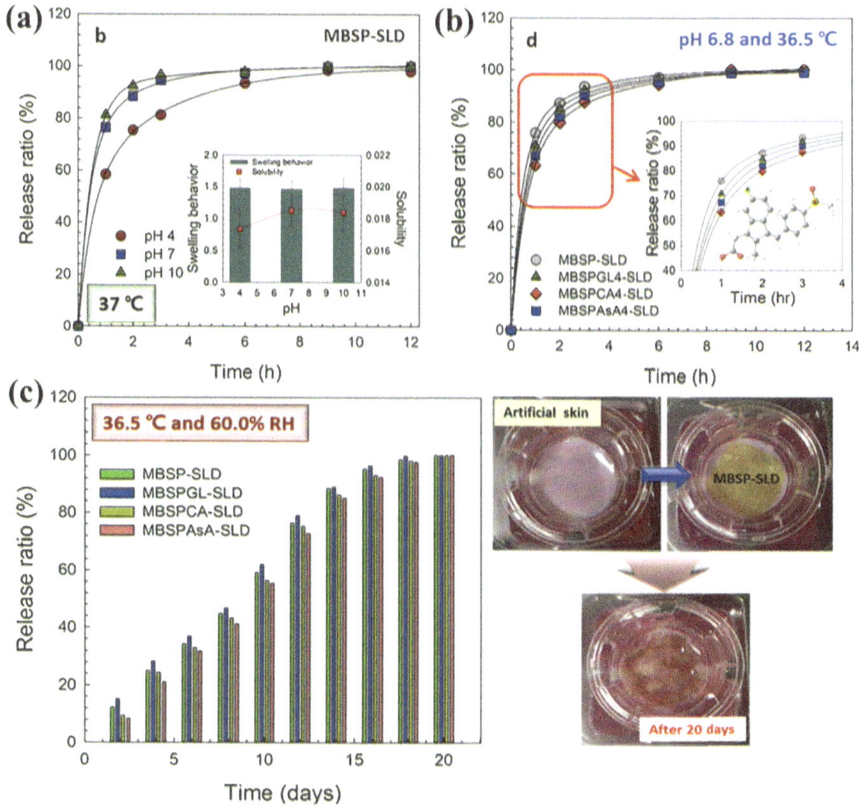

Fig. 6.8 SLD release ratio (%) on MBS/PVA biomaterial imprinted films. (**a**) The effect of different pH on release rate at 37 °C; (**b**) SLD release ratio (%) of imprinting films with different plasticizers (pH 6.8 and 37 °C); (**c**) SLD release ratio (%) and imagings of imprinted films at artificial skin (60.0% RH and 36.5 °C). (Reproduced with permission of [29])

6.5.1 Construction of Reservoir Transdermal Patch

The manufacture of the reservoir-type transdermal patch included three steps: synthesis of MIP composite membrane, construction of gel reservoir and assembly of the transdermal patch.

- **Synthesis of MIP composite membrane.** Firstly, the composite membrane was synthesized by incubating Acetobacter xylinum in coconut juice, and then immersed in 10% 3-MPS-toluene solution (w/w) to activate for 5 h at 80 °C. The membrane thickness was 5 μm, the resistance was 1 Ω cm^2, and the tensile strength was 3 kN m^{-2}. Next, the pre-polymerization solution was prepared by dissolving *S*-propranolol HCl, MAA, EDMA and AIBN in DMF. After degassing under vacuum and purging with N_2, the polymerization solution was poured on the surface of the activated cellulose membrane and reacted at 60 °C for 20 h.

Then, the imprinted membrane was rinsed with 500 mL methanol/acetic acid (90/10, v/v) and methanol to clean the remaining template. The NIP membrane was synthesized by using the same preparation process in the absence of template. Finally, the thickness of the obtained MIP and NIP composite film was about 6 μm, the resistance was 2 Ω cm^2, and the tensile strength was 8 kN m^{-2}.

- **Construction of gel reservoir.** Firstly, dissolving 2.5 g of chitosan in 100 mL of 10% acetic acid solution (w/w), and adding 10% NaOH solution (w/w) dropwise until a white precipitate was produced. After incubating for 12 h in 37 °C oven, the chitosan gel solution was rinsed three times with 5 mL distilled water, then stood overnight to remove residual bubbles. Finally, racemic propranolol hydrochloride was added and mixed uniformly to form a drug reservoir. The final concentration was 300 μg mL^{-1}.
- **Assembly of the transdermal patch.** The gel reservoir-type transdermal patch was composed of a backing layer, gel reservoir, MIP film, and release liner with the 16 cm^2 effective surface area. Racemic propranolol hydrochloride (1.5 mg) was dissolved into chitosan gel solution (0.5 mL) to obtain the reservoir formulation. Then, a 20 cm^2 MIP membrane and an adhesive-coated release liner were laminated together to form an "adhesive-membrane laminate." The empty patch device was formed by attaching a 1.5 cm^3 spacer and an "adhesive-membrane laminate" together. The spacer was obtained from the Scotchpak™ foam tape with the thickness of 1.0 mm and internal dimension of 4 cm. A 0.5 mL gel reservoir was poured into the separator, and then the separator was glued together with the backing laminate by epoxy resin. Finally, a complete reservoir transdermal patch was formed.

6.5.2 Evaluation of the Transdermal Patch

The vertical Franz diffusion cells was used to investigate the in vitro transdermal permeability. The MIP, NIP, or cellulose membrane was adhered to the epidermal side of excised Wistar rat dorsal skin. The skin was fixed on the donor compartment with the surface of dermis side skin contacting the receptor solution. Then, 300 μg/mL of racemic propranolol gel (0.5 mL) was coated onto the donor compartment. The receiving cell contained 25 mL of PBS buffer (pH 7.4), and the skin surface temperature was maintained at 32 °C. 250 μL of sample solution was taken out at regular intervals to analyze drug content and immediately an equal volume of buffer solution was added. The results indicated that the pure propranolol gel reservoir showed no difference in the stereoselective transdermal permeation, the J_{ss} of *S*-isomer and *R*-isomer were 0.20 ± 0.02 and 0.23 ± 0.03 μg cm^{-2} h^{-1}, respectively (Fig. 6.9a). In contrast, the stereoselective transdermal permeation of propranolol on the MIP composite membrane showed a significant difference. The J_{ss} of *S*-isomer was 0.33 ± 0.05 μg cm^{-2} h^{-1}, while *R*-isomer was only 0.11 ± 0.01 μg cm^{-2} h^{-1} (Fig. 6.9b). The mean ratio (*S*/*R*) of percutaneous permeation on the MIP composite membranes was 3.01 ± 0.42, which was significantly higher than that of the

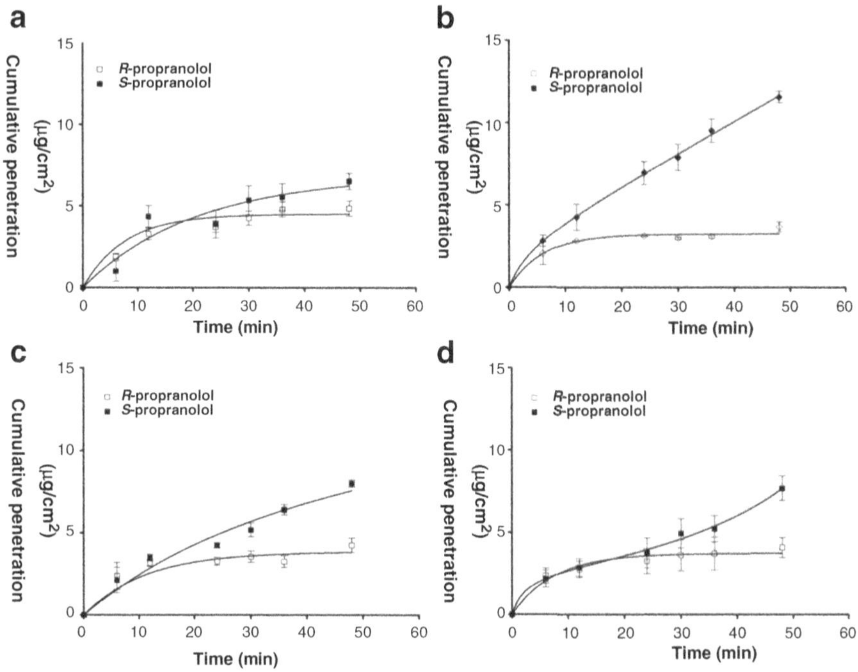

Fig. 6.9 Permeation curves of *R*/*S*-propranolol enantiomers from gels reservoir on rat skin alone (**a**), MIP membrane-skin (**b**), NIP membrane-skin (**c**) and cellulose membranes-skin (**d**). (Reproduced with permission of [34])

cellulose membranes ($S/R = 1.73 \pm 0.25$) and NIP composite membranes ($S/R = 1.54 \pm 0.12$). These results indicated that the MIP composite membrane can stereoselectively control the transdermal delivery of *S*-propranolol when the gel reservoir was loaded with racemic propranolol.

In the in vivo evaluation experiment, rats were anesthetized with 30% carbamate (0.9 g/kg, intra-abdominal) and their back hair was removed. Prior to being attached to the dorsal region of the rat, the transdermal patch was peeled off the paper liner. The gel reservoir loading with 1.5 mg racemic propranolol or 0.75 mg *S*-propranolol as control patch was fixed to the dorsal skins for transdermal penetration experiment. The backing films attached to spacers had the function of protecting gel loss and preventing water evaporation. At 0 h and every 3–6 h, a 250 μL blood sample was drawn from the femoral vein and implanted in a heparin tube within 24 h. After centrifuged at $3000 \times g$ for 10 min, the obtained plasma was stored in a vial at −20 °C until use. Then, 50 μL of 40 ng mL^{-1} *S*-phenylephrine aqueous solution (internal standard) was added to 125 μL of plasma sample, and the sample was extracted with 3 mL of ether. After the organic solvent of extract was evaporated, the resultant residue was dissolved in 125 μL of mobile phase and analyzed by chiral HPLC to obtain the concentration of propranolol.

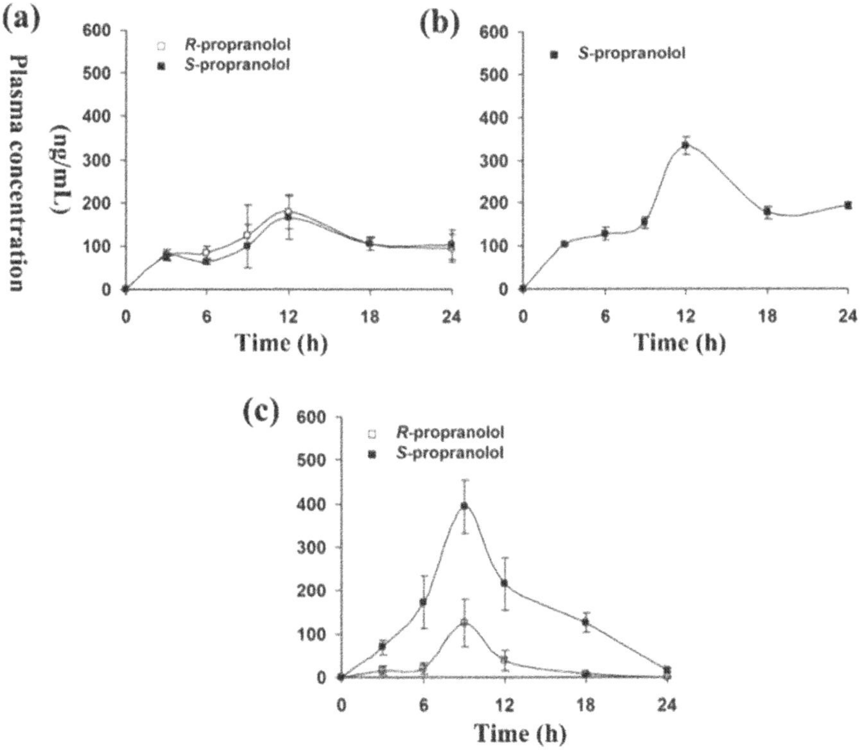

Fig. 6.10 Plasma concentration versus time curves in vivo evaluation experiment for gel reservoir-type transdermal patch: (**a**) gel formulation containing 1.5 mg racemic propranolol, (**b**) gel formulation containing 0.75 mg *S*-propranolol, (**c**) the MIP patch containing 1.5 mg racemic propranolol (mean ± SE, $n = 3$). (Reproduced with permission of [34])

Figure 6.10 showed the relationship between plasma concentration and time after transdermal application of the racemic propranolol-MIP patch and topical racemic propranolol or *S*-propranolol gel formulation. As shown in Fig. 6.10c, after the application of the MIP patch, the plasma concentration of the *S*-propranolol enantiomer reached the maximum at the eighth hour ($T_{max} = 8$ h). In contrast, after using topical racemic propranolol or *S*-propranolol gel formulation, the T_{max} of *S*-propranolol was about 12 h, indicating lower absorption efficiency (Fig. 6.10a,b). In the MIP patch system, the C_{max} and AUC_{0-24} of the *S*-propranolol enantiomer were approximately three and five times than that of the *R*-propranolol enantiomer, respectively. However, the racemic propranolol gel formulation showed no enantiomeric differences in terms of C_{max} and AUC_{0-24}. There was no significant difference of the *S*-propranolol enantiomer concentration in plasma when using the MIP patches or gel formulations containing only pure *S*-propranolol. The above results indicated that the release of the *S*-isomer of propranolol could be selectively regulated via the MIP patch. In addition, mechanistic studies had shown that the MIP membrane selectively controlled the *S*-isomer release by forming a composite with selective receptor sites, while *R*-enantiomer was released by a route without selectivity. Based on the above discussion, the reservoir transdermal patch prepared by the MIP composite membrane might have great potential for stereoselective control of transdermal delivery.

6.6 The Limitations and Outlook

The comfortable, low first-pass metabolism and high bioavailability of transdermal patches combined with the high drug loading, low toxicity and controlled release property of MIP provides a potential formulation type for the development of various types of transdermal delivery system. Although MIP-based transdermal patches have made some progress, there are still some limitations or deficiencies that need to be further improved, such as skin irritation due to the use of some acidic monomers, lack of actual clinical verification. Most studies just stay in the proof-of-concept stage, almost no actual pharmaceutical preparations have been formed. In addition to the MIP microspheres and MIP membrane-based transdermal delivery devices that have been developed, more functional MIP materials, such as photoresponsive, temperature-sensitive, and restricted access MIP, would provide new opportunities for the development of smarter transdermal delivery devices.

References

1. Prausnitz MR, Langer R (2008) Transdermal drug delivery. Nat Biotechnol 26:1261–1268
2. Prausnitz MR, Mitragotri S, Langer R (2004) Current status and future potential of transdermal drug delivery. Nat Rev 3:115–124
3. Wang CY, Javadi A, Ghaffari M, Gong SQ (2010) A pH-sensitive molecularly imprinted nanospheres/hydrogel composite as a coating for implantable biosensors. Biomaterials 31:4944–4951
4. Canfarotta F, Lezina L, Guerreiro A, Czulak J, Petukhov A, Daks A, Smolinska-Kempisty K, Poma A, Piletsky S, Barlev NA (2018) Specific drug delivery to cancer cells with double-imprinted nanoparticles against epidermal growth factor receptor. Nano Lett 18:4641–4646
5. Hwang I, Kim HN, Seong M, Lee SH, Kang MS, Yi H, Bae WG, Kwak MK, Jeong HE (2018) Multifunctional smart skin adhesive patches for advanced health care. Adv Healthc Mater 1800275:1–20
6. Lulinski P (2017) Molecularly imprinted polymers based drug delivery devices: a way to application in modern pharmacotherapy. A review. Mater Sci Eng 76:1344–1353
7. Benowitz NL, Hukannen J, Jacob IP (2009) Nicotine chemistry, metabolism, kinetics and biomarkers. Handbook Exp Pharm 192:29–60
8. White HK, Levin ED (2004) Chronic transdermal nicotine patch treatment effects on cognitive performance in age-associated memory impairment. Psychopharmacology 171:465–471
9. Ruela ALM, Figueiredo EC, Pereira GR (2014) Molecularly imprinted polymers as nicotine transdermal delivery systems. Chem Eng J 248:1–8
10. Olivier JC, Rabouan S, Couet W (2003) In vitro comparative studies of two marketed transdermal nicotine delivery systems: Nicopatch® and Nicorette®. Int J Pharm 252:133–140

11. Ruelaa ALM, Figueiredob EC, Carvalhob FC, Araújob MB, Pereira GR (2018) Adsorption and release of nicotine from imprinted particles synthesised by precipitation polymerisation: optimising transdermal formulations. Eur Polym J 100:67–76
12. Ruela ALM, Figueiredo EC, Perissinato AG, Lima ACZ, Araújo MB, Pereira GR (2013) In vitro evaluation of transdermal nicotine delivery systems commercially available in Brazil. Braz J Pharm Sci 49:579–588
13. Yoshikawa M, Tharpa K, Dima ŞO (2016) Molecularly imprinted membranes: past, present, and future. Chem Rev 116:11500–11528
14. Ciardelli G, Borrelli C, Silvestri D, Cristallini C, Barbani N, Giusti P (2006) Supported imprinted nanospheres for the selective recognition of cholesterol. Biosens Bioelectron 21:2329–2338
15. Hattori K, Hiwatari M, Iiyama C, Yoshimi Y, Kohori F, Sakai K, Piletsky SA (2004) Gate effect of theophylline-imprinted polymer grafted to the cellulose by living radical polymerization. J Membr Sci 233:169–173
16. Ulbricht M, Belter M, Langenhangen U, Schneider F, Weigel W (2002) Novel molecularly imprinted polymer (MIP) composite membranes via controlled surface and pore functionalizations. Desalination 149:293–295
17. Lehmann M, Brunner H, Tovar GEM (2002) Selective separations and hydrodynamic studies: a new approach using molecularly imprinted nanosphere composite membranes. Desalination 149:315–321
18. Chronakis IS, Jakob A, Hagström B, Ye L (2006) Encapsulation and selective recognition of molecularly imprinted theophylline and 17β-estradiol nanoparticles within electrospun polymer nanofibers. Langmuir 22:8960–8965
19. Yoshikawa M, Murakoshi K, Kogita T, Hanaoka K, Guiver MD, Robertson GP (2006) Chiral separation membranes from modified polysulfone having myrtenal-derived terpenoid side groups. Eur Polym J 42:2532–2539
20. Ramamoorthy M, Ulbricht M (2003) Molecular imprinting of cellulose acetate-sulfonated polysulfone blend membranes for rhodamine B by phase inversion technique. J Membr Sci 217:207–214
21. Klemm D, Schumann D, Udhardt U, Marsch S (2001) Bacterial synthesized cellulose-artificial blood vessels for microsurgery. Prog Polym Sci 26:1561–1603
22. Fontana JD, De-Souza AM, Fontana CK, Torriani IL, Moreschi JC, Gallotti BJ, De-Souza SJ, Narcisco GP, Bichara JA, Farah LFX (1990) Acetobacter cellulose pellicle as a temporary skin substitute. Appl Biochem Biotechnol 25:253–264
23. Bodhibukkana C, Srichana T, Kaewnopparat S, Tangthong N, Bouking P, Martin GP, Suedee R (2006) Composite membrane of bacterially-derived cellulose and molecularly imprinted polymer for use as a transdermal enantioselective controlled-release system of racemic propranolol. J Control Release 113:43–56
24. Jantarat C, Tangthong N, Songkro S, Martin GP, Suedee R (2008) S-propranolol imprinted polymer nanoparticle-on-microsphere composite porous cellulose membrane for the enantioselectively controlled delivery of racemic propranolol. Int J Pharm 349:212–225
25. Fu L, Zhu J, Zhang S, Li X, Zhang B, Pu H (2018) Hierarchical structure and thermal behavior of hydrophobic starch-based films with different amylose contents. Carbohydr Polym 181:528–535
26. Ali A, Xie F, Yu L, Liu H, Meng L, Khalid S (2018) Preparation and characterization of starch-based composite films reinforced by polysaccharide-based crystals. Compos Part B Eng 133:122–128
27. Lam PL, Gambari R (2014) Advanced progress of microencapsulation technologies: in vivo and in vitro models for studying oral and transdermal drug deliveries. J Control Release 178:25–45
28. Kwak MK, Jeong HE, Suh KY (2011) Rational design and enhanced biocompatibility of a dry adhesive medical skin patch. Adv Mater 23:3949–3953
29. Tak HY, Yun YH, Lee CM, Yoon SD (2019) Sulindac imprinted mungbean starch/PVA biomaterial films as a transdermal drug delivery patch. Carbohydr Polym 208:261–268

30. Noori S, Kokabi M, Hassan ZM (2019) Poly(vinyl alcohol)/chitosan/honey/clay responsive nanocomposite hydrogel wound dressing. J Appl Polym Sci 135:46311–46322
31. Sánchez-González EG, Yépez-Mulia L, Hernández-Abad VJ, Cook HJ (2015) The influence of polymorphism on the manufacturability and in vitro dissolution of sulindac-containing hard gelatin capsules. Pharm Dev Technol 20:306–313
32. Ravi Kumar MNV (2000) A review of chitin and chitosan applications. React Funct Polym 46:1–27
33. Chandy T, Sharma CP (1993) Chitosan matrix for oral sustained delivery of ampicillin. Biomaterials 14:939–944
34. Suedee R, Bodhibukkana C, Tangthong N, Amnuaikit C, Kaewnopparat S, Srichana T (2008) Development of a reservoir-type transdermal enantioselective-controlled delivery system for racemic propranolol using a molecularly imprinted polymer composite membrane. J Control Release 129:170–178

Chapter 7
MIP as Drug Delivery Systems of Anticancer Agents

Jing Feng and Zhaosheng Liu

7.1 Active Targeting

For the treatment of cancer, the traditional method of administration lacks specificity for tumor cells, and even a therapeutic dose of drugs will produce certain toxic and side effects on normal cells, which make treatment difficult. In order to avoid the above defects, a targeted drug delivery system (TODDS) has been proposed [1–3]. TODDS is the selective concentration of drugs through local administration, gastrointestinal tract or systemic blood circulation by means of carriers, ligands, or antibodies to target tissues, target organs, target cells, or intracellular structures [4]. TODDS has four advantages: (1) can selectively kill cells, (2) can kill chemotherapy-insensitive or drug-resistant cells, (3) targeted drugs can play a synergistic role with chemotherapy drugs, (4) the toxicity of targeted drugs is relatively low [4–6]. After being designed and constructed reasonably and effectively, TODDS can deliver drugs effectively to the tumor site for the purpose of specific targeted delivery of drugs, reduce the distribution of drugs in normal tissues, the toxic and side effects of drugs, and improve the bioavailability of drugs greatly [4, 7].

TODDS is mainly divided into passive targeting and active targeting [8]. Active targeting, also known as ligand-mediated targeting can be achieved by modifying ligands such as antibodies, nucleic acid aptamers, peptides, polysaccharides, and some special small molecules to the surface of the polymer, thereby making the polymer to be captured specifically by tumor cells overexpressing the corresponding receptor [8–10]. The method of targeting molecules is an important mechanism of endocytosis. After the polymer enters the tumor cell, it is endocytosed by the tumor cell through the targeting molecule [10].

J. Feng · Z. Liu (✉)
College of Pharmacy, Tianjin Medical University, Tianjin, China

Z. Liu et al. (eds.), *Molecularly Imprinted Polymers as Advanced Drug Delivery Systems*, https://doi.org/10.1007/978-981-16-0227-6_7

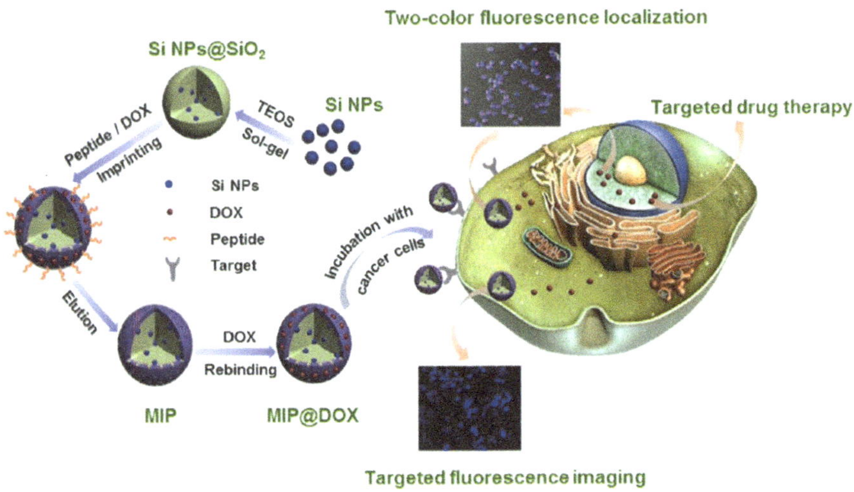

Scheme 7.1 Illustration for preparation of MIP for targeted fluorescence imaging and targeted therapy in the cancer cell. (Reproduced with permission of [12])

Active targeting mainly includes four types. The first type is bio-oriented type. It is through grafting ligands on the surface of drug carriers to carry antitumor drugs into tumor cells. Relying on the specific recognition between ligands and receptors, drug carriers can accumulate on the surface of cells in large amounts to cause intracellular change [11]. This type can increase the number of intracellular vectors greatly.

Wang et al. [12] used linear peptides from the extracellular domain of human epidermal growth factor receptor 2 (HER2) and doxorubicin (DOX) as templates to prepare fluorescent silicon nanoparticles (SiNPs) coated with dual-template molecularly imprinted polymer (MIP). Due to the imprinted sites of peptides formed on the surface of nanoparticles, the MIP could target breast cancer cells selectively and specifically, performed fluorescence imaging at the cellular level, and the DOX-loaded MIP (MIP@DOX) could be used as a therapeutic probe for targeting and killing breast cancer effectively.

In order to prove that the imprinting site of the MIP which had the targeting effect of overexpression on the cell surface in combination with HER2 protein (Scheme 7.1), the nanoparticles were applied to SK-BR-3 cells for fluorescence imaging. The results are shown in Figs. 7.1 and 7.2. The cells are able to take up more MIPs, compared with non-imprinted polymer (NIP), showing a greater signal intensity, and the fluorescent signal increases with time significantly. In contrast, the fluorescent signal from the NIP in SK-BR-3 cells is weaker than the corresponding MIP at each time point. Figure 7.2 shows the normalized average fluorescence intensity (MFI) of the MIP and NIP fluorescent silicon nanoparticles cultured with SK-BR-3 cells, and the MIP is 1.86 times than that of the NIP. The MIPs which adsorbed different

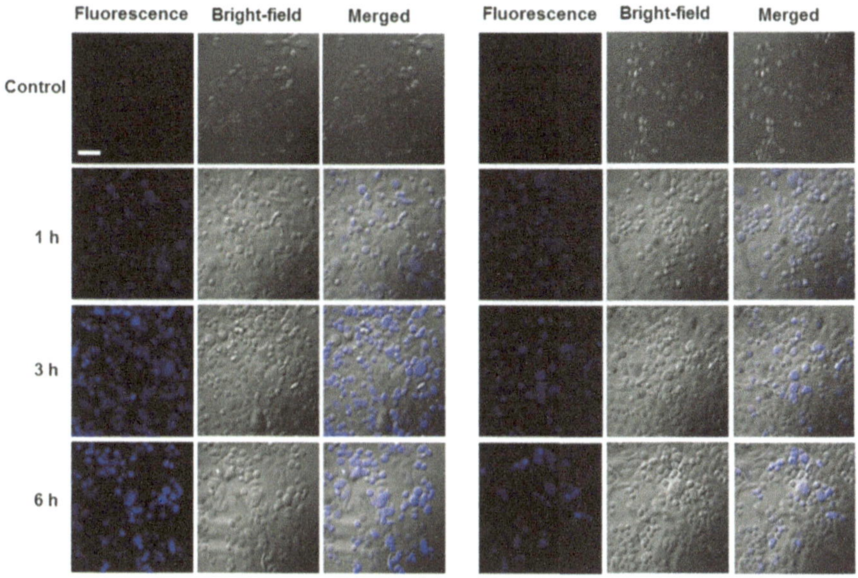

Fig. 7.1 Confocal fluorescence images of SK-BR-3 cells after incubating with 40 μg mL^{-1} of MIP and NIP at certain time points (1, 3, and 6 h). (Reproduced with permission of [12])

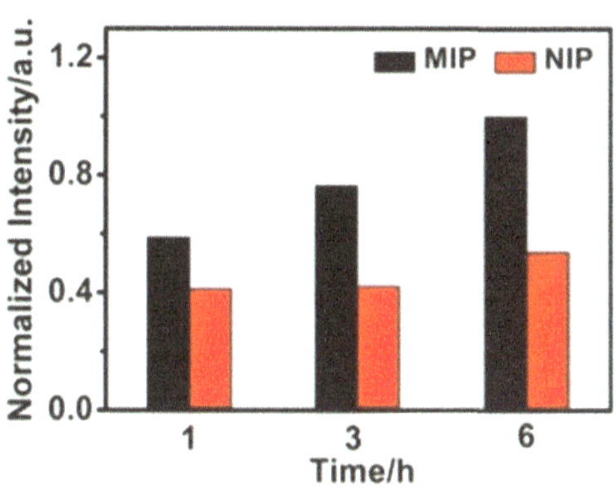

Fig. 7.2 Normalization of mean fluorescence intensity (MFI) of SiNPs from MIP and NIP incubated with SK-BR-3 cells at 1, 3 and 6 h. (Reproduced with permission of [12])

amounts of peptides were incubated with BT-474 cells 6 h (Fig. 7.3). The latter treated cells showing weaker fluorescence intensity because more targeting sites on the MIP were occupied by peptides. The above results all confirmed the active targeting ability of the MIP to HER2-positive cells.

Dong et al. [13] proposed a strategy of using HER2 *N*-glycan nanomolecularly imprinted polymers (nanoMIPs) to target the treatment of HER2-positive breast cancer. Scheme 7.2 illustrates the principle of the nanoMIP targeting HER2 and

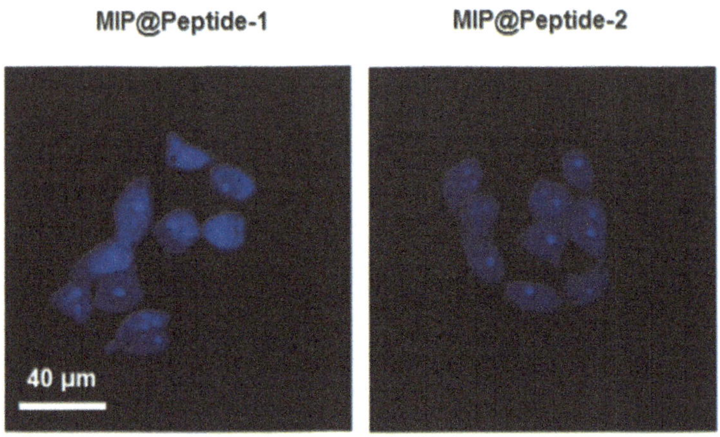

Fig. 7.3 Confocal fluorescence pictures of BT-474 cells incubated with 40 μg mL^{-1} of MIP@Peptide-1 and MIP@Peptide-2 (peptide in MIP@Peptide-1 was 30.9 mg $g^{-1}$and in MIP@Peptide-2 was 52.6 mg g^{-1}) for 6 h. (Reproduced with permission of [12])

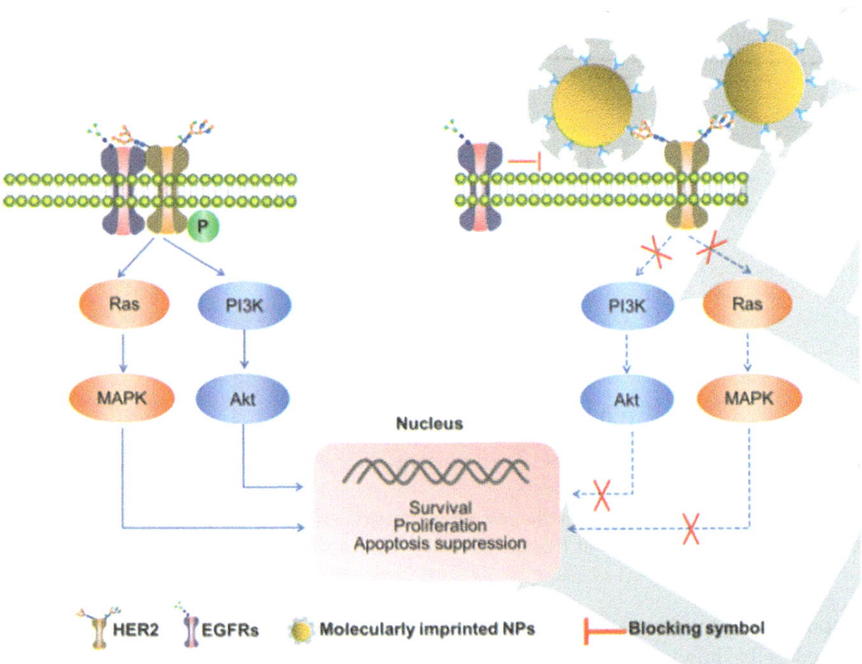

Scheme 7.2 Illustration of the principle of blocking the HER2 signaling pathway via HER2-glycan-imprinted nanoparticles. (Reproduced with permission of [13])

inhibiting the growth of HER2-positive breast cancer. HER2 *N*-glycan imprinted nanoMIPs can bind to almost all HER2 glycans. Once bound to HER2, the glycan imprinted MIP will prevent the dimerization of HER2 with other members due to steric hindrance, thereby causing downstream signaling pathways to inhibit the growth of HER2-positive breast cancer.

The second type is internal stimulation targeting, which means that the carrier has higher stability or structural integrity in normal tissue cells. After reaching the tumor cells, due to changes in the surrounding environment, its stability is reduced, and its structure damaged, thereby releasing the drug carried at the tumor site. Internal stimuli often includes pH, temperature, etc. The most typical example is to change pH for achieving targeted drug delivery. The specific description can be seen in Sect. 7.2.

The third type is external stimulus targeting. It mainly uses the physical and chemical properties of the drug carrier itself. When the external environment changes, the physical and chemical properties of the carrier will also change to a certain extent, thereby releasing the carried anti-tumor drugs to achieve the purpose of targeting actively to the tumor site. External stimuli includes infrared light, magnetism, etc. The specific description can be seen in Sect. 7.3.

The last type is the multi-function type, mainly through the way of embedding, coupling to assemble the above-mentioned several types of targeted functional materials to create a multifunctional drug delivery system for the purpose of maximizing the advantages of intelligence, reducing patient adverse reactions [14].

Liu et al. [15] prepared a kind of sialic acid (SA) imprinted nanoparticles, using the *cis*-diol structure of SA and phenylboronic acid (PBA) groups to undergo an esterification reaction, which was affected by pH control, only in an acidic environment [16, 17], PBA could form borate with SA and further released NO, thereby having a targeting effect on tumor cells. In order to verify the targeting ability of *S*-nitrosothiol MIP, the survival rates of cells with high SA expression levels (MCF-7 cells and HepG2 cells) and low SA expression levels (HeLa cells and normal hepatocytes) were measured. The results showed that the cell survival rate of cells with high SA expression level increased with the increasing of the concentration of the *S*-nitrosothiol MIP; while the survival rate of cells with low SA expression level did not change significantly. This showed the targeting properties of the SA imprinted nanoparticles to tumor cells.

7.2 pH-Controlled Anticancer Drug Release with the Aid of MIP

In addition to the active targeting of cancer cells, MIP can also control the release of anticancer drugs through pH. The drug release properties of MIP mainly depend on the bond between the template and monomer to form MIP, including hydrogen bond, coordination bond, van der Waals force, and electrostatic interaction [18, 19]. The use of pH to adjust the release is based on the pH value in the cell [20]. Because the

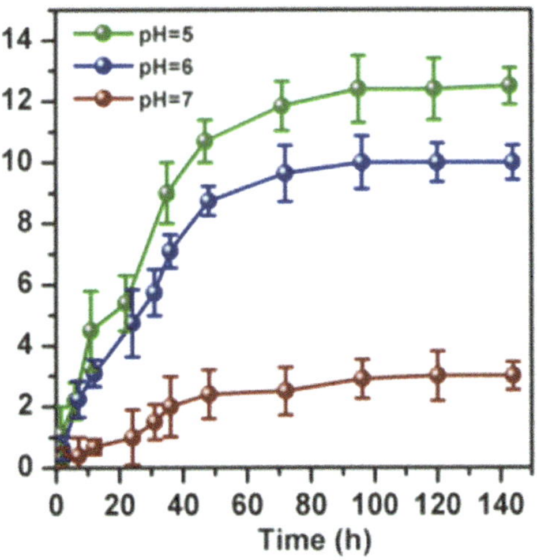

Fig. 7.4 Drug release of PTX in phosphate buffer saline (pH = 5, 6 and 7). (Reproduced with permission of [22])

pH value of cancer cells is higher than that of normal cells (pH = 5.8) [19–21], a pH-controlled DDS can be designed to achieve the selective release of anticancer drugs in cancer cells. This function has been widely used in the delivery of anticancer drugs.

Bai et al. [22] used paclitaxel (PTX) as a template to synthesize a novel MIP particles, containing polymethacrylic oligomeric silsesquioxane (M-POSS) by RAFT precipitation polymerization (RAFTPP), to achieve the controlled release of PTX at different pH (Fig. 7.4). At pH 7, the drug releasing is slow, and only 3.1% is released within 50 h. On the contrary, at pH 5 and 6, the drug release rate is accelerated significantly. This might be due to the hydrogen bond between the template and the monomer breaking under weakly acidic conditions, causing the drug to separate. This provides a feasible way to effectively kill tumor cells.

Zhang et al. [23] prepared a surface molecularly imprinted polymer (SMIP) using DOX as a template on the surface of mesoporous silica nanoparticles (MSNs) to achieve pH-controlled drug release (Fig. 7.5). It was found that at pH 5.0, the release rate of DOX was faster than that at pH 7.4, which might be due to the acidic conditions that may help the degradation of the polymer by breaking the hydrogen bond between DOX and the monomer, so the release amounts could be as high as 50%. These results indicated that the prepared SMIP was an effective nanocarrier capable of acting on cancer cells.

Zhang et al. [24] used the coordination bonds of metal ion to mediate the interaction between the template and functional monomer to synthesize a doxorubicin-molecularly imprinted hydrogel (MIH) capable of pH response and continuous drug release. A hydrogel is a hydrophilic polymer gel between solid and liquid that can swell in water but cannot dissolve in water [25, 26], which consists of a three-dimensional network or interpenetrating network structure composed of polymer chains and swells in water to thermodynamic equilibrium [26–28]. Conventional hydrogels have poor release effects due to defects such as drug

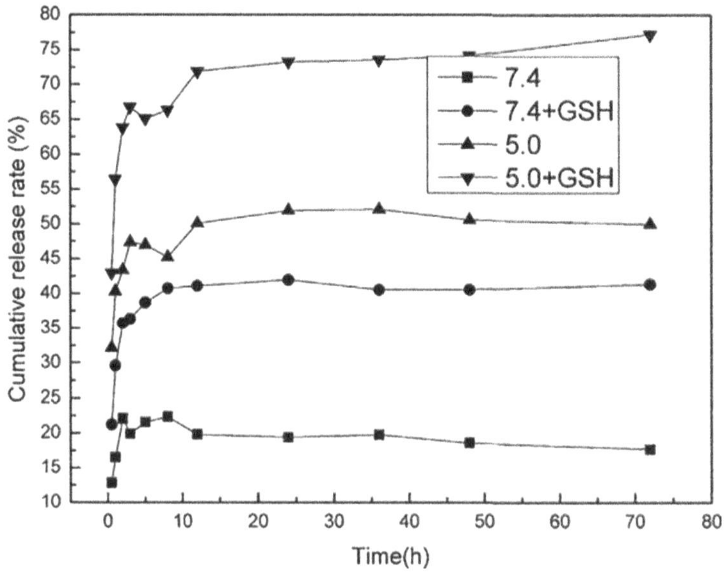

Fig. 7.5 Release curves of DOX from the nanocarrier under different conditions. (Reproduced with permission of [23])

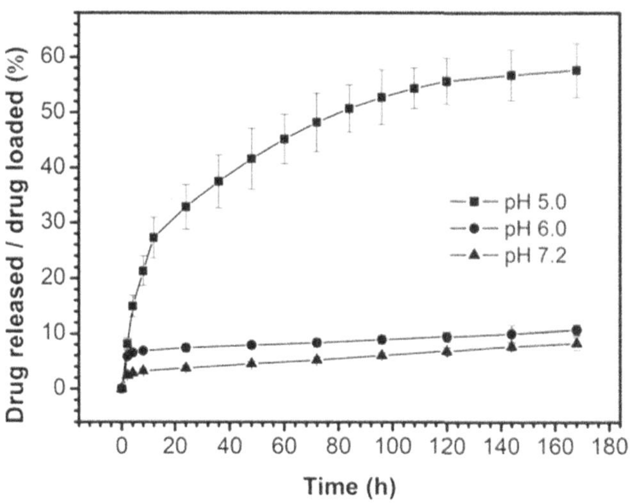

Fig. 7.6 Release profiles of DOX from the imprinted hydrogel at different pH values. (Reproduced with permission of [24])

diffusion and polymer degradation, so MIH is needed to be designed to overcome this shortcoming.

The release of MIH in vitro was studied in different pH release media (Fig. 7.6). Within 7 days, MIH showed significant pH control. When the pH was 5.0, the drug release was the fastest, capable of releasing 60% of the load, while at pH of 7.2 or 6.0, only less than 10% of the drug was released. The result of this drug release was due to the interaction between the copper ion and DOX template and 4-vinylpyridine (4-VP) functional monomer. According to previous reports [29], when the pH was in the alkaline or neutral range, 4-VP could deprotonate and chelate stably with copper ions. As the pH value decreased, the protonated form of 4-VP increased, leading to the breaking of the coordination bond between 4-VP and copper ions, and at the same time, the DOX and copper ion imprinting sites also be dissociated.

7.3 Stimuli-Responsive Imprinted DDS for Cancer Treatment

The molecularly imprinted polymers can not only release drugs under the action of pH, they can also deliver drugs in response to other stimuli. In order to avoid the early release of antitumor drugs during the circulation in the body, and at the same time overcome the fact that the drug carrier reaches the tumor tissue and enters the tumor cell and cannot release the anticancer drug effectively, the DDS is usually designed according to the different physiological environment of the tumor tissue and tumor cells. It can respond to the environment in the body to achieve a more effective drug delivery or be designed to respond to external stimuli at tumor cells. Therefore, a stimuli-responsive polymer (SRP) is favorable. SRP can be regarded as a smart molecule, which can cause changes in chemical structure or phase state based on external signals [30, 31]. This change usually manifests as a huge change in the physical and chemical properties of the polymer. The external signal can be chemical factors such as redox, enzyme concentration, and pH, which can also be the fluctuations of the body, such as temperature, light and heat, magnetic field, and ultrasound [32].

One of the applications of SRP is as a carrier for drug delivery [33, 34]. In general, they can self-assemble into nano- or micro-carriers with specific structures and dimensions [33–35]. The common morphology is spherical micelles or hollow vesicles. The polymer assembles drug molecules internally in a covalent or non-covalent manner, and then further realizes the drug release of the intelligent carrier [36]. Therefore, the drug release parameters of MIP can be controlled by adjusting external signals.

Stimulus response drug delivery system (SRPs-DDS) is a combination of SRP and DDS, and is a functional DDS. The stimulation methods of SRPs-DDS are mainly divided into three types: chemical stimulation, biological stimulation, and physical stimulation [37, 38]. The type of the developed SRPs-DDS mainly includes chemical stimulus response, biological stimulus response, temperature response and light response. Stimulated response imprinting DDS is a kind of intelligent drug release, which refers to the release of drugs according to specific environmental stimuli (temperature, light and heat, infrared, magnetic field, etc.) to extend the duration of pharmacological effects.

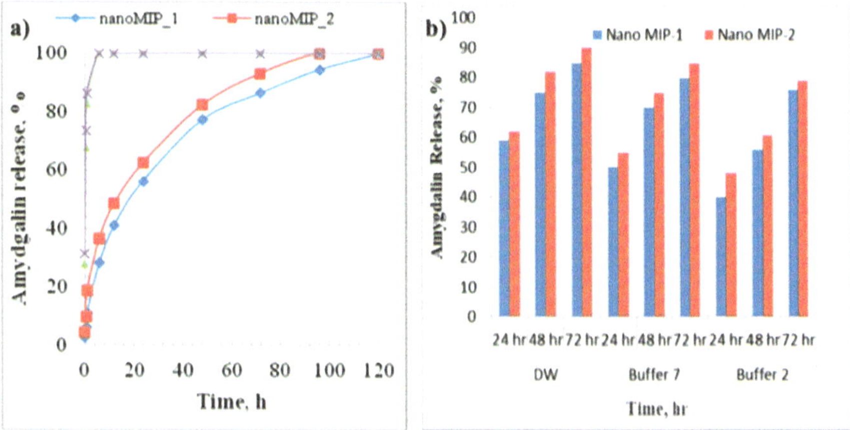

Fig. 7.7 (**a**) Amygdalin release from nanoMIPs and control polymers (NIPs). (**b**) Release of amygdalin from nanoMIPs in different medium. (Reproduced with permission of [43])

Suksuwan et al. [39] designed and synthesized heat-responsive MIP nanoparticles with the selectivity to (*R*)-thalidomide, and drugs were delivered to tumor cells successfully. Thalidomide (α-phthalimide glutamine) is a sedative agent introduced by the German pharmaceutical company, Grande in the 1950s [40]. It is a glutamic acid derivative and an inhibitor of angiogenesis, and an effective inhibitor of tumor necrosis factor-α (TNF-α) with *S* and *R* enantiomers [41, 42]. Methacrylic acid, 2,6-bis (acrylamido) pyridine (as the functional monomer), *N*,*N*′-methylene-bisacrylamide (as the cross-linking agent), 2,2′-azobis (2-methylbutyronitrile) (as a free radical initiator) were used to prepare grafting MIP, resulting in a polymer network with precisely designed binding sites. The polyacrylamide is proven to have dual stimulating properties in water, which can lead to a synergistic response of nanoparticles. Poloxamers are composed of triblock copolymers of polyethylene oxide (PEO) and polypropylene oxide (PPO) blocks. It has high biocompatibility, good drug resistance of tumor cells due to the hydrophobicity of PPO.

Korde et al. [43] prepared amygdalin nanoMIPs with high drug loading and thermal response controlled release. Amygdalin is a natural product with antitumor activity, small side effects, wide sources, and relatively low price. It can be decomposed to produce hydrocyanic acid through the action of β-D-glucosidase, thereby killing tumor cells [44, 45]. The release of amygdalin from the nanoMIPs was studied in an in vitro cell experiment. Figure 7.7a shows the release of the NIPs and nanoMIPs within 7 days, and Fig. 7.7b shows the release of the nanoMIPs loaded with drugs in deionized water, buffer 7 and 2. Compared with the nanoMIP, the release of amygdalin from the NIPs was faster, and the release could complete within 60 min; while the amounts of amygdalin released by the nanoMIP within 24 h was 60%, 50%, and 35%, respectively. This result might be due to the swelling of

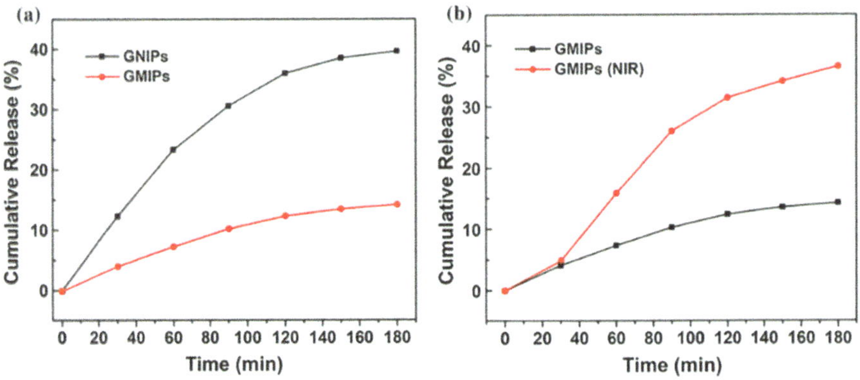

Fig. 7.8 (**a**) Kinetic release curves of DOX from GNIPs to GMIPs; (**b**) the cumulative release of DOX from GMIPs under or without the irradiation of 808-nm laser (2 W/cm^2) for different time. (Reproduced with permission of [46])

the nanoMIPs in the release medium, which slowed down the drug release; while the presence of non-specific binding sites in the nanoNIPs made the release rate faster.

Xu et al. [46] prepared graphene quantum dot (GQD) doped DOX@MIP through microemulsion polymerization to achieve controlled release of DOX in response to near-infrared (NIR). GQD has NIR absorption capacity, strong photothermal conversion efficiency and excellent thermal conductivity in the body [47]. At the same time, their high surface area and various surface functional groups make it possible to load drugs [47–49]. In vitro release experiments were performed in PBS buffer at pH 7.4 (Fig. 7.8a). GNIPs were initially released quickly (10–60 min), then slowly, and the drug release rate exceeded 23.34% within 60 min. For the GMIPs, the release was very slow, especially 10–60 min, and only 180.21% was released in 180 min (Fig. 7.8b). But then, the cumulative release of DOX increased almost linearly to 26.02% at 90 min, and increased slowly to 36.54% at 180 min. The controlled release results indicated that the GQD in the GMIPs microspheres converted the laser energy of 808 nm into heat, which raised the temperature, resulting in changes in the binding energy between the graphene quantum dots and the drug [50].

Zhu et al. [51] prepared vinblastine (VBL)-loaded polymer nanoparticles (VBL-NPs) by molecular imprinting. Using acryloyl amino acid comonomer and disulfide bond cross-linking agent to obtain VBL-NPs-PEG-FA polymer, and then using the redox reactivity of disulfide bonds, delayed release of VBL can be achieved.

Griffete et al. [52] prepared an alternative magnetic nanoparticles response to magnetic field (AMF) that could control the release of DOX in vitro and living cells. They designed a new magnetic DOX delivery system (Fe_2O_3@DOX-MIP), by using DOX as a template on the surface of Fe_2O_3 nanoparticles to synthesize the SMIP. Then under the condition of AMF, the interaction between the drug and the template was destroyed, thereby triggering the release of DOX.

7.4 Biocompatibility and Biodegradability of MIP-DDS

Biocompatibility is a concept that the interaction between biomedical materials and the human body produces various complex biological, physical and chemical reactions [53], i.e., the ability of living tissues to react to inactive materials, and generally refers to the compatibility between the material and the host [54]. It often includes two categories of histocompatibility and hemocompatibility: histocompatibility—refers to cell adhesion, non-inhibitory cell growth, cell activation, resistance to cytoplasmic transformation, resistance to disease, and non-antigenicity, non-mutagenic, non-carcinogenic, non-teratogenic, etc.; blood compatibility—refers to antiplatelet thrombosis, anticoagulation, anti-hemolysis, anti-leukopenia, anticomplement system hyperactivity, anti-plasma protein adsorption, and anti-cytokine adsorption [55]. After biomaterials are implanted in the human body, they will have an effect on the environment of specific biological tissues, and biological tissues will also have an impact on biological materials. Medical materials should have good biocompatibility, nontoxic to the human body, non-pyrogen, no irritation to surrounding tissues, no hemolytic reaction after implantation, no immune rejection reaction, no carcinogenic and teratogenic effects, can be degraded and absorbed and replaced by collective organizations gradually.

Bai et al. [22] prepared a molecularly imprinted polymer using PTX as a template by precipitation polymerization. For PTX, PTX-loaded particles, and PTX-free particles were tested for cytotoxicity in the HepG2 cell line. The results are shown in Fig. 7.9. The cell survival rate of particles without PTX load is higher, even at the lowest concentration, the cell survival rate can still reach 80%, which shows that the MIP does not have any cytotoxicity to the HepG2 cell line and has good biocompatibility. Figure 7.9b shows that compared with PTX drugs, the PTX-loaded microparticles have higher cytotoxicity at different concentrations.

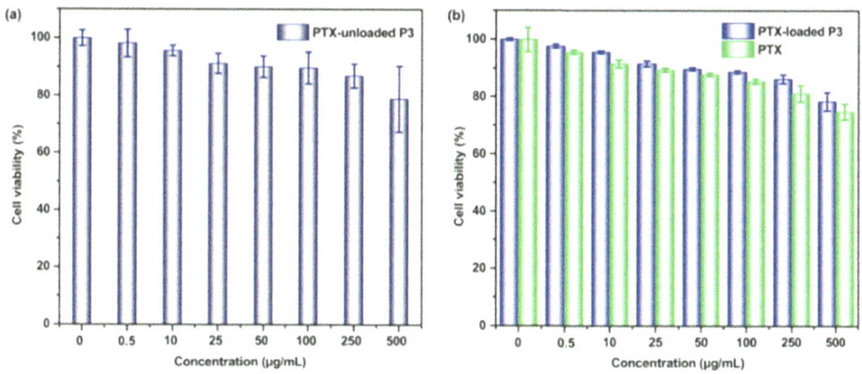

Fig. 7.9 Cytotoxic effect of pure PTX, PTX-unloaded particles, and PTX-loaded particles incubated with HepG2. (Reproduced with permission of [22])

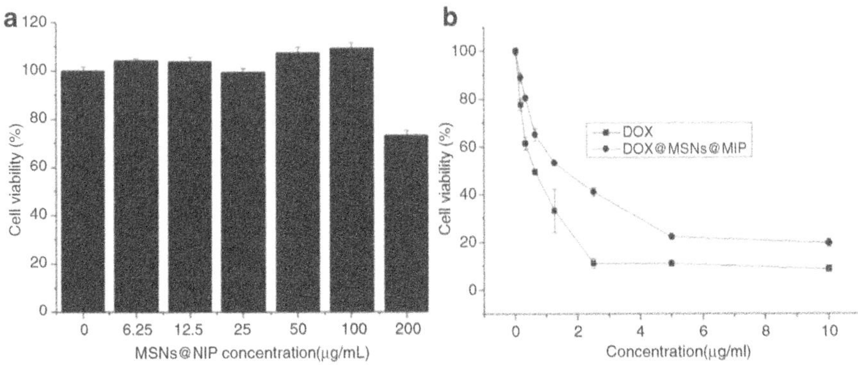

Fig. 7.10 Effect of MSNs@NIP (**a**), DOX@MSNs@MIP and DOX (**b**) on viability of Tca8113 cells after 24 h culture. (Reproduced with permission of [23])

Zhang et al. [23] used metal ion coordination bonds to mediate the interaction between the template and the monomer to synthesize a DOX-MIH capable of pH response and continuous drug delivery. As shown in Fig. 7.10a, the MSN@NIP were no cytotoxicity to TCA8113 cells (cell viability near 100%) in the concentration range 0–10 μg/mL, which indicated that it had good biocompatibility. When the DOX concentration was 2.5 μg/mL, the cell death was close to 60% for the DOX@MSNs@MIP and close to 89% for DOX. When the concentration increased to 5 μg/mL, the cell death reached 80% for the DOX@MSNs@MIP and close to 90% for DOX. This result verified the invasion of the DOX within SMIP into TCA8113 cancer cells and indicated that the prepared SMIP was an effective nanocarrier.

Canfarotta et al. [56] synthesized MIPs and performed toxicological evaluation of such NPs in different cell lines (HaCaT, MEF, HT1080, and macrophages). The biocompatibility of the imprinted NPs was evaluated in HaCaT, HT1080 and MEF cells. HaCaT and HT1080 belong to keratinocyte cell line and fibrosarcoma cell line, respectively. MEF cells have high respiratory activity, and its phosphorylation rate is highly responsive to any changes [57, 58]. As shown in Fig. 7.11, three concentrations of the imprinted NP (50, 25 and 10 μg/mL) are incubated with HaCaT and HT1080 cells for 24 and 48 h for MTT test. The test results showed that for the naked NP, PEG1100 coated MIP or PEG4000 coated, the cell survival rate of MIP did not decrease significantly, indicating that the imprinted nanoparticles had no obvious cytotoxicity and good cell compatibility.

The imprinted nanoparticles were further evaluated by measuring oxidative phosphorylation (ATP), oxygen consumption rate (OCR) and extracellular acidification (ECA) on MFF cells. As shown in Fig. 7.12, the cell activity of naked NP, PEG1100-coated MIP or PEG4000-coated MIP did not change significantly, proving that the imprinted nanoparticles were not affected. The oxygen consumption rate of MEF cells treated with PEG1100-coated MIP decreased by 57%, while the

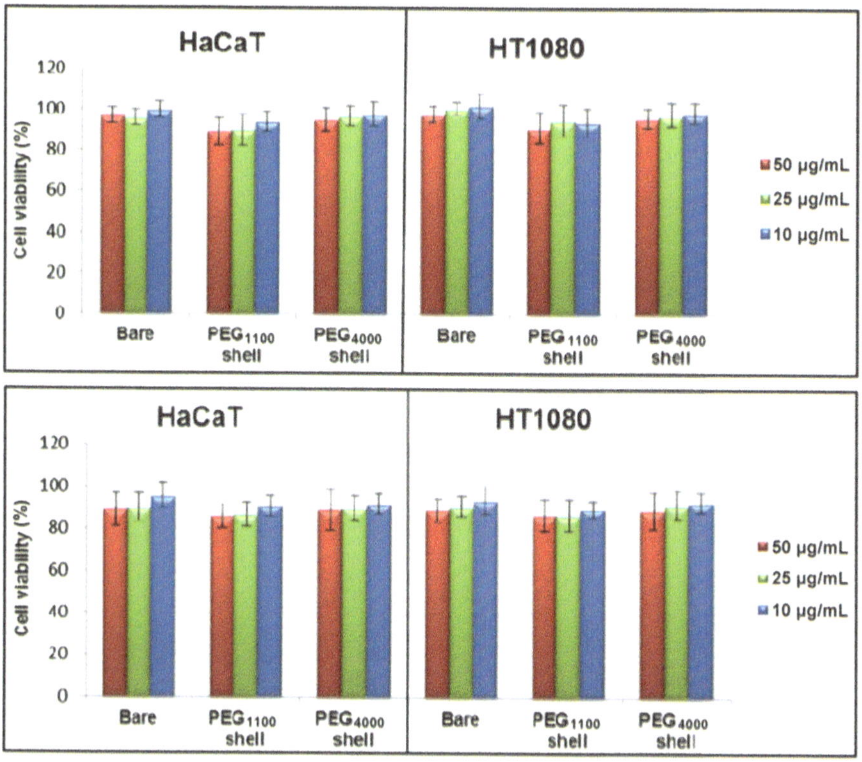

Fig. 7.11 MTT test performed on both HaCaT and HT1080 cells after 24 and 48 h. (Reproduced with permission of [56])

acidification level increased; while the oxygen consumption of MEF cells treated with PEG4000-coated MIPs decreased by 22%, but the acidification level was almost unchanged.

7.5 In Vivo Evaluation of Imprinted DDS

Hashemi-Moghaddam et al. [59] used dopamine as monomer, HER2 conformation epitope and DOX as template to synthesize MIP on the surface of SiNPs. The resulting MIP was used to treat ovarian cancer for mouse model, and DOX targeting delivery was realized. The tumor volume trend after treatment is shown in the Fig. 7.13. The tumors of the control group and the non-medicated epitope group increased significantly over time, while the tumor growth of the DOX and DOX-IP group was lower than that of the control group and the epitope group significantly. Compared with all other groups, the tumor volume in the DOX-EPI-IP treatment

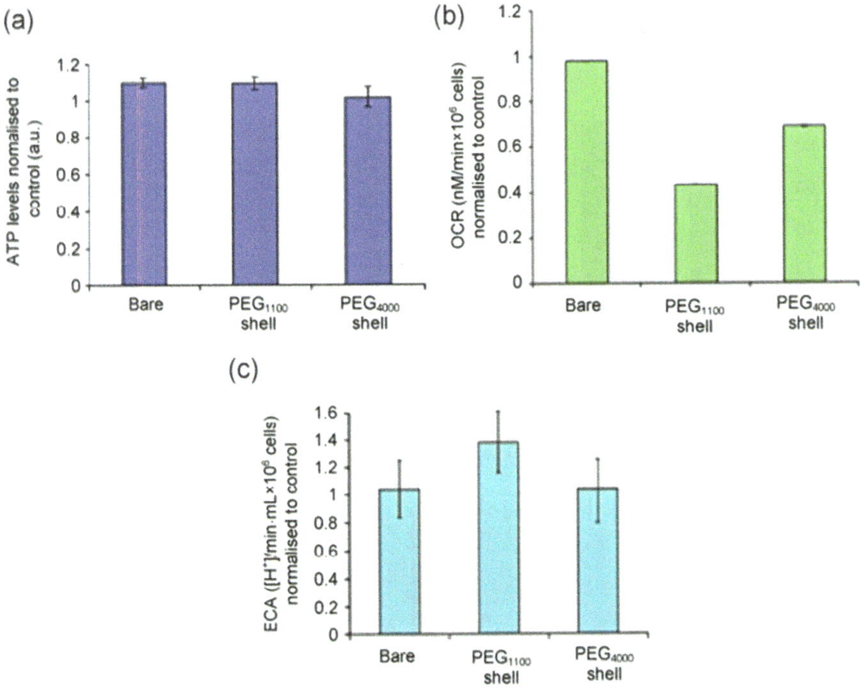

Fig. 7.12 ATP levels in MEF cells after 24-h exposure to a 50 μg/mL solution of molecularly imprinted polymers (MIPs) (**a**), OCR (**b**), and ECA (**c**) assays performed on the same sample. (Reproduced with permission of [56])

group was significantly reduced, indicating that the DOX-EPI-IP group had the best tumor growth inhibition effect. Figure 7.14 shows the tumor inhibition rate (IR) of different treatment methods. Compared with all other groups, IR increased significantly after DOX-EPI-IP treatment, with the highest IR% value. In contrast, in DOX and DOX-IP groups, IR increased from day 3 to day 10, and then started decrease, indicating that it only inhibited the tumor temporarily.

As shown in Table 7.1, the DT and SGR of tumor volume between the experimental groups have significant differences. The DT of the DOX EPI-IP treatment group was 6.037 days; it was higher than other experimental groups significantly; the SGR of the DOX EPI-IP treatment group was 0.11, which was lower than all other experimental groups significantly. Tumors treated with free DOX showed tumor growth DT and SGR of 4.42 days and 0.15, respectively, which were lower and higher than those of the DOX-EPI-IP treatment group, respectively.

Dong et al. [13] proposed the use of HER2 *N*-glycan imprinted nanoMIPs to treat HER2-positive breast cancer. In vivo experiments studied the biodistribution of nanoMIPs in nude mice by fluorescence imaging (Fig. 7.15). Figure 7.15a shows the biodistribution after intravenous injection of nanoMIP and nanoNIP to mice at 1, 5, 9, 13 and 15 days. For the MIP, it could be observed that there was a strong

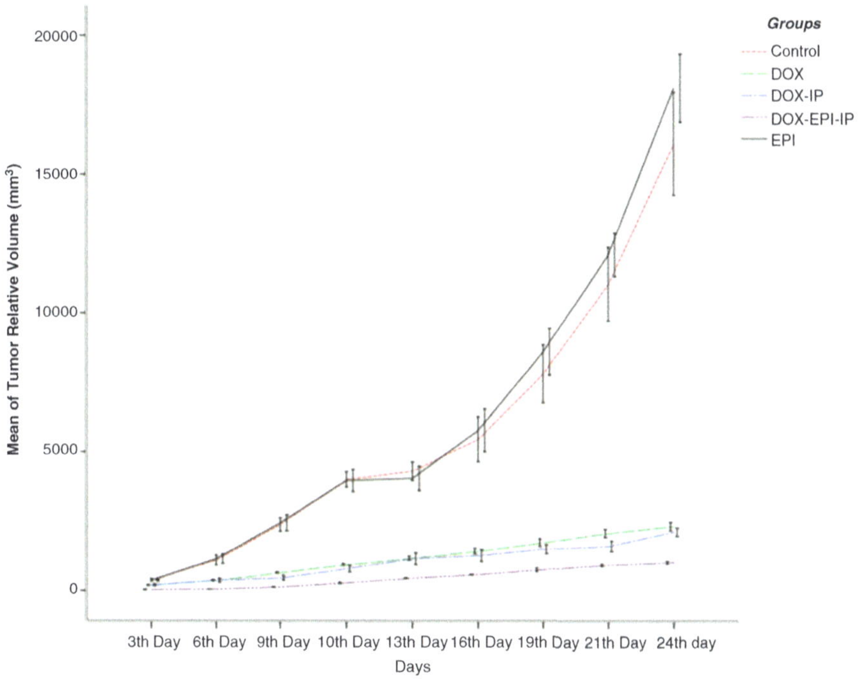

Fig. 7.13 Relative tumor volume of tumor-bearing mice that received different treatments. (*EPI* epitope, *DOX* doxorubicin, *DOX-IP* doxorubicin-imprinted polymer, *DOX-EPI-IP* doxorubicin–epitope-imprinted polymer). (Reproduced with permission of [59])

fluorescent signal at the tumor site, and the signal intensity increased with the increase of days. In contrast, for the nanoNIP, almost no fluorescent signal could be observed after 5 days, which showed that the nanoMIP has a specific recognition effect on tumors. Figure 7.15b–d shows the changes in tumor volume, weight and size with the number of days of injection. The PBS and NIPs treatment groups showed rapid tumor growth, while significant tumor growth inhibition was observed in the mice treated with the nanoMIPs. At the same time, the nanoMIPs hardly caused weight changes in mice during the treatment period (Fig. 7.15e), which indicated that nanoMIPs had almost no biological toxicity.

7.6 Conclusion

In summary, this chapter outlines the drug delivery system using MIP as an anticancer agent. MIP can achieve the active targeting of drugs through three methods: bio-directed, internal stimulated, and multifunctional, so that tumor cells can further capture polymers specifically; it can also achieve control release of

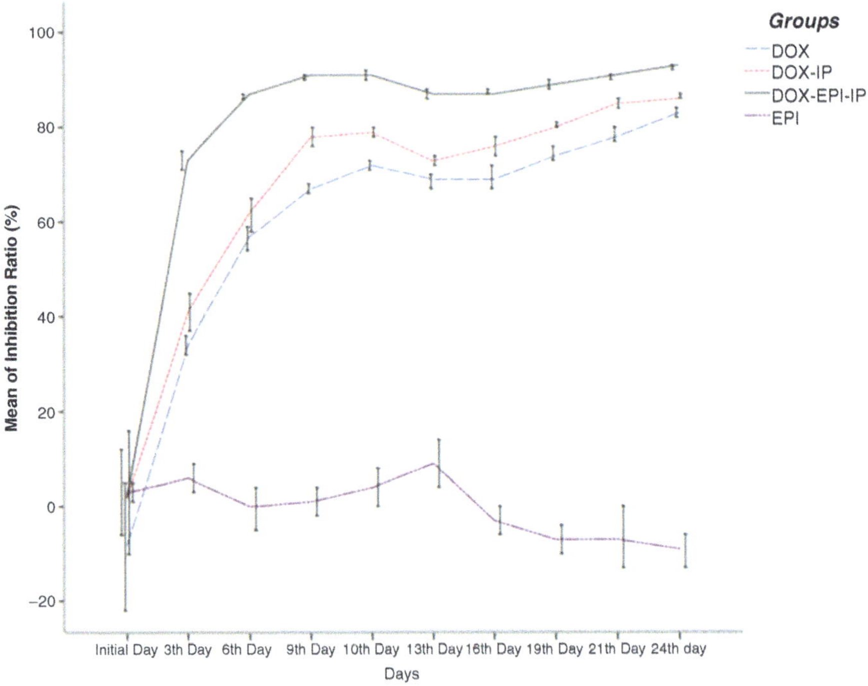

Fig. 7.14 Tumor growth inhibition ratio (IR%) of different treatment groups. (*EPI* epitope, *DOX* doxorubicin, *DOX-IP* doxorubicin-imprinted polymer, *DOX-EPI-IP* doxorubicin–epitope-imprinted polymer). (Reproduced with permission of [59])

Table 7.1 Parameters of tumor growth in experimental groups. (Reproduced with permission of [59])

Groups	DT	SGR
Control	2.871 ± 0.112^a	0.242 ± 0.009^a
DOX	4.428 ± 0.142^b	0.157 ± 0.005^b
DOX-EPI-IP	6.037 ± 0.172^c	0.115 ± 0.003^c
DOX-IP	4.283 ± 0.159^b	0.162 ± 0.006^b
EPI	2.800 ± 0.064^a	0.248 ± 0.006^a

Different superscript letters in the same columns indicate significant differences ($p < 0.05$)

anticancer agents under specific stimuli such as biological and physical based on the different environments of cancer cells and normal cells. Tests have proved that MIP has good biocompatibility in the body and can achieve biodegradation. The above results illustrate the feasibility of MIP as a drug delivery system for anticancer agents, and it is expected to be used in further cancer treatments.

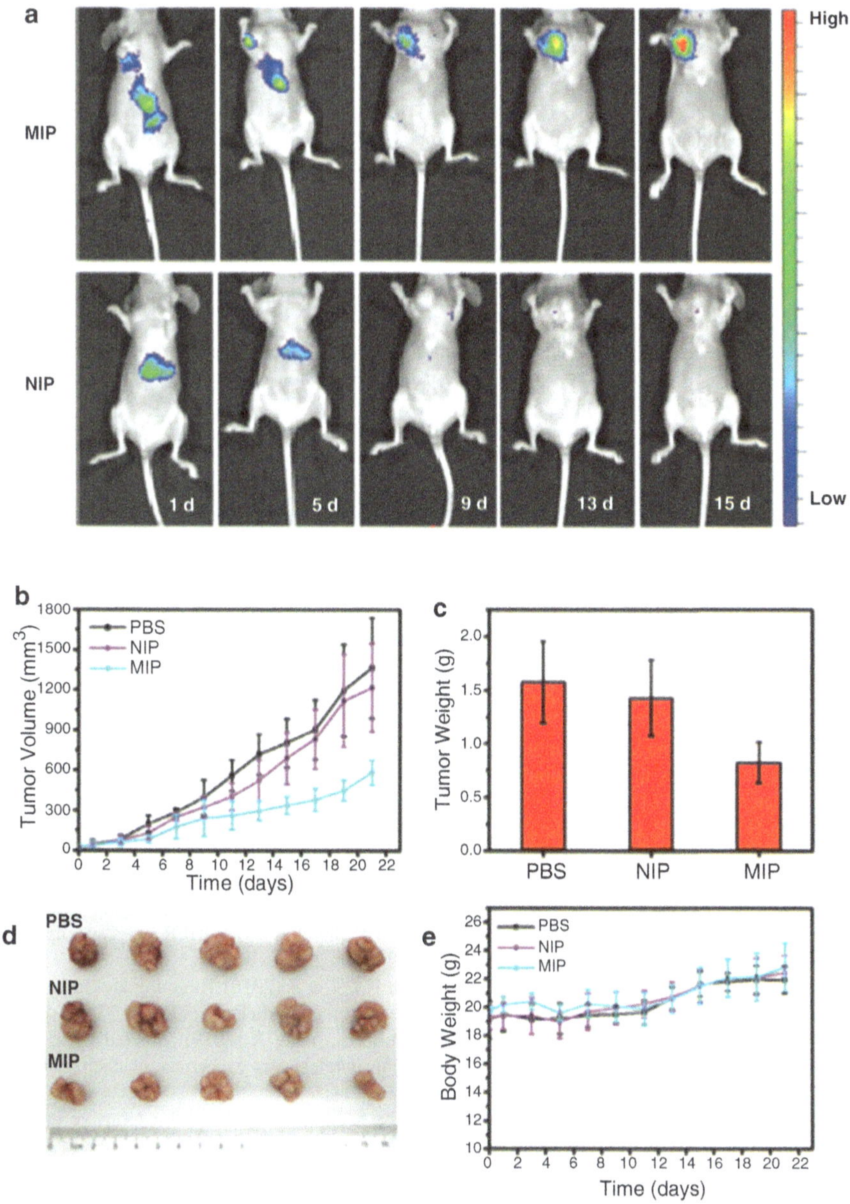

Fig. 7.15 (**a**) In vivo fluorescence imaging of SKBR-3 tumor after intravenous injection of HER2-glycan-imprinted, NIR797-doped SiO_2 NPs (top) or non-imprinted, NIR797-doped SiO_2 NPs (bottom). (**b**) Mean tumor volume. (**c**) Mean tumor weights after excision at day 21. (**d**) Representative photographs of mice tumors after treatment for 21 days. (**e**) Body weight of the mice in different groups after treatment at different time intervals. (Reproduced with permission of [13])

References

1. Hema S, Thambiraj S, Shankaran DR (2018) Nanoformulations for targeted drug delivery to prostate cancer: an overview. JNN 18:5171–5191
2. Vivek R, Thangam R, Nipunbabu V, Ponraj T, Kannan S (2014) Oxaliplatin-chitosan nanoparticles induced intrinsic apoptotic signaling pathway: a "smart" drug delivery system to breast cancer cell therapy. Int J Biol Macromol 65:289–297
3. Maeda H, Tominaga K, Iwanaga K, Nagao F, Habu M, Tsujisawa T, Seta Y, Toyoshima K, Fukuda J, Nishihara T (2009) Targeted drug delivery system for oral cancer therapy using sonoporation. J Oral Pathol Med 38:572–579
4. Masood F (2016) Polymeric nanoparticles for targeted drug delivery system for cancer therapy. Mater Sci Eng C Mater Biol Appl 60:569–578
5. Goldstein D, Nassar T, Lambert G, Kadouche J, Benita S (2005) The design and evaluation of a novel targeted drug delivery system using cationic emulsion–antibody conjugates. J Control Release 108:418–432
6. Taratula O, Dani RK, Schumann C, Xu H, Wang A, Song H, Dhagat P, Taratula O (2013) Multifunctional nanomedicine platform for concurrent delivery of chemotherapeutic drugs and mild hyperthermia to ovarian cancer cells. Int J Pharm 458:169–180
7. Krukiewicz K, Zak JK (2016) Biomaterial-based regional chemotherapy: local anticancer drug delivery to enhance chemotherapy and minimize its side-effects. Mater Sci Eng C Mater Biol Appl 62:927–942
8. Fathi S, Oyelere AK (2016) Liposomal drug delivery systems for targeted cancer therapy: is active targeting the best choice. Future Med Chem 8:2091–2112
9. Sutton D, Nasongkla N, Blanco E, Gao J (2007) Functionalized micellar systems for cancer targeted drug delivery. Pharm Res 24:1029–1046
10. Dai L, Liu J, Luo Z, Li M, Cai K (2016) Tumor therapy: targeted drug delivery systems. J Mater Chem B 4:6758–6772
11. Tatematsu K, Iijima M, Yoshimoto N, Nakai T, Okajima T, Kuroda S (2016) Bio-nanocapsules displaying various immunoglobulins as an active targeting-based drug delivery system. Acta Biomater 35:238–247
12. Wang HY, Cao PP, He ZY, He XW, Li WY, Li YH, Zhang YK (2019) Targeted imaging and targeted therapy of breast cancer cells via fluorescent double template-imprinted polymer coated silicon nanoparticles by an epitope approach. Nanoscale 11:17018–17030
13. Dong Y, Li W, Gu Z, Xing R, Ma Y, Zhang Q, Liu Z (2019) Inhibition of HER2-positive breast cancer growth by blocking the HER2 signaling pathway with HER2-glycan-imprinted nanoparticles. Angew Chem Int Ed Engl 58:10621–10625
14. Wang S, Yin D, Wang W, Shen X, Zhu JJ, Chen HY, Liu Z (2016) Targeting and imaging of cancer cells via monosaccharide-imprinted fluorescent nanoparticles. Sci Rep 6:22757–22768
15. Liu T, Qiao Z, Wang J, Zhang P, Zhang Z, Guo DS, Yang X (2019) Molecular imprinted S-nitrosothiols nanoparticles for nitric oxide control release as cancer target chemotherapy. Colloids Surf B: Biointerfaces 173:356–365
16. Liu H, Li Y, Sun K, Fan J, Zhang P, Meng J, Wang S, Jiang L (2013) Dual-responsive surfaces modified with phenylboronic acid-containing polymer brush to reversibly capture and release cancer cells. J Am Chem Soc 135:7603–7609
17. Bull SD, Davidson MG, Van den Elsen JM, Fossey JS, Jenkins AT, Jiang YB, Kubo Y, Marken F, Sakurai K, Zhao J, James TD (2013) Exploiting the reversible covalent bonding of boronic acids: recognition, sensing, and assembly. Acc Chem Res 46:312–326
18. Gupta P, Vermani K, Garg S (2002) Hydrogels: from controlled release to pH-responsive drug delivery. Drug Discov Today 7:569–579
19. Song SW, Hidajat K, Kawi S (2007) PH-controllable drug release using hydrogel encapsulated mesoporous silica. Chem Commun 42:4396–4398

20. Kamada H, Tsutsumi Y, Yoshioka Y, Yamamoto Y, Kodaira H, Tsunoda S, Okamoto T, Mukai Y, Shibata H, Nakagawa S, Mayumi T (2004) Design of a pH-sensitive polymeric carrier for drug release and its application in cancer therapy. Clin Cancer Res 10:2545–2550
21. Alvarez-Bautista A, Duarte CMM, Mendizabal E, Katime I (2016) Controlled delivery of drugs through smart pH-sensitive nanohydrogels for anti-cancer therapies: synthesis, drug release and cellular studies. Des Monomers Polym 19:319–329
22. Bai J, Zhang Y, Chen L, Yan H, Zhang C, Liu L, Xu X (2018) Synthesis and characterization of paclitaxel-imprinted microparticles for controlled release of an anticancer drug. Mater Sci Eng C Mater Biol Appl 92:338–348
23. Zhang K, Guan X, Qiu Y, Wang D, Zhang X, Zhang H (2016) A pH/glutathione double responsive drug delivery system using molecular imprint technique for drug loading. Appl Surf Sci 389:1208–1213
24. Zhang Q, Zhang L, Wang P, Du S (2014) Coordinate bonding strategy for molecularly imprinted hydrogels: toward pH-responsive doxorubicin delivery. J Pharm Sci 103:643–651
25. Abbasi F, Mirzadeh H (2010) Properties of poly (dimethyl siloxane)/hydrogel multicomponent systems. J Polym Sci Part B Polym Phys 41:2145–2156
26. Malkoch M, Vestberg R, Gupta N, Mespouille L, Dubois P, Mason AF, Hedrick JL, Liao Q, Frank CW, Kingsbury K, Hawker CJ (2006) Synthesis of well-defined hydrogel networks using click chemistry. Chem Commun 26:2774–2776
27. Huang G, Gao J, Hu Z, John JV, Ponder BC, Moro D (2004) Controlled drug release from hydrogel nanoparticle networks. J Control Release 94:303–311
28. Luo Y, Shoichet MS (2004) A photolabile hydrogel for guided three-dimensional cell growth and migration. Nat Mater 3:249–253
29. Li SH, Wang J, Zhao MP (2009) Cupric ion enhanced molecular imprinting of bovine serum albumin in hydrogel. J Sep Sci 32:3359–3363
30. Stuart MA, Huck WT, Genzer J, Müller M, Ober C, Stamm M, Sukhorukov GB, Szleifer I, Tsukruk VV, Urban M, Winnik F, Zauscher S, Luzinov I, Minko S (2010) Emerging applications of stimuli-responsive polymer materials. Nat Mater 9:101–113
31. Liu F, Urban MW (2010) Recent advances and challenges in designing stimuli-responsive polymers. Prog Polym Sci 35:3–23
32. Calderón M, Quadir MA, Strumia M, Haag R (2010) Functional dendritic polymer architectures as stimuli-responsive nanocarriers. Biochimie 92:1242–1251
33. Ninawe PR, Parulekar SJ (2012) Drug delivery using stimuli-responsive polymer gel spheres. Ind Eng Chem Res 51:1741–1755
34. Yang MY, Tan L, Wu HX, Liu CJ, Zhuo RX (2015) Dual-stimuli-responsive polymer-coated mesoporous silica nanoparticles used for controlled drug delivery. J Appl Polym Sci 132:42395–42404
35. Peng X, Wei L, Jing X, Cui L, Wu J, Meng G, Liu Z, Guo X (2018) Stimuli-responsive nano-polymer composite materials based on the triazine skeleton structure used in drug delivery. JOM 71:308–314
36. Qiu X, Hu S (2013) "Smart" materials based on cellulose: a review of the preparations, properties, and applications. Materials 6:738–781
37. Unsoy G, Gunduz U (2018) Smart drug delivery systems in cancer therapy. Curr Drug Targets 19:202–212
38. Jérme C (2010) Macromolecular engineering and stimulus response in the design of advanced drug delivery systems. MRS Bull 35:665–672
39. Suksuwan A, Lomlim L, Rungrotmongkol T, Nakpheng T, Dickert FL, Suedee R (2015) The composite nanomaterials containing (R)-thalidomide-molecularly imprinted polymers as a recognition system for enantioselective-controlled release and targeted drug delivery. J Appl Polym Sci 132:41930–41951
40. Franks ME, Macpherson GR, Figg WD (2004) Thalidomide. Lancet 363:1802–1811
41. D'Amato R (2000) Methods and compositions for inhibition of angiogenesis. Biotechnol Adv 15:779–779

42. Tursen B, Tursen U (2014) Treatment options in Behcet's disease. Global J Dermatol 2:27–49
43. Korde BA, Mankar JS, Phule S, Krupadam RJ (2019) Nanoporous imprinted polymers (nanoMIPs) for controlled release of cancer drug. Mater Sci Eng C Mater Biol Appl 99:222–230
44. Song Z, Xu X (2014) Advanced research on anti-tumor effects of amygdalin. J Cancer Res Ther 10:3–7
45. Kaminskia BM, Steinhilber D, Stein JM, Ulrich S (2012) Phytochemicals resveratrol and sulforaphane as potential agents for enhancing the anti-tumor activities of conventional cancer therapies. Curr Pharm Biotechnol 13:137–146
46. Xu Y, Hu X, Guan P, Du C, Tian Y, Ding S, Li Z, Yan C (2019) A novel controllable molecularly imprinted drug delivery system based on the photothermal effect of graphene oxide quantum dots. J Mater Sci 54:9124–9139
47. Bacon M, Bradley SJ, Nann T (2014) Graphene quantum dots. Part Part Syst Charact 1:415–428
48. Wang X, Sun X, Lao J, He H, Cheng T, Wang M, Wang S, Huang F (2014) Multifunctional graphene quantum dots for simultaneous targeted cellular imaging and drug delivery. Colloids Surf B Biointerfaces 122:638–644
49. Wang Z, Xia J, Zhou C, Via B, Xia Y, Zhang F, Tang J (2013) Synthesis of strongly green-photoluminescent graphene quantum dots for drug carrier. Colloids Surf B Biointerfaces 112:192–196
50. Kim H, Lee D, Kim J, Kim TI, Kim WJ (2013) Photothermally triggered cytosolic drug delivery via disruption using a functionalized reduced graphene oxide. ACS Nano 7:6735–6746
51. Zhu Y, Liu R, Huang H, Zhu Q (2019) Vinblastine-loaded nanoparticles with enhanced tumor-targeting efficiency and decreasing toxicity: developed by one-step molecular imprinting process. Mol Pharm 16:2675–2689
52. Griffete N, Fresnais J, Espinosa A, Wilhelm C, Bée A, Ménager C (2015) Design of magnetic molecularly imprinted polymer nanoparticles for controlled release of doxorubicin under an alternative magnetic field in athermal conditions. Nanoscale 7:18891–18896
53. Learmonth ID (2003) Biocompatibility: a biomechanical and biological concept in total hip replacement. Surgeon 1:1–8
54. Lofti M, Nejb M, Naceur M (2013) Chapter 8: Cell adhesion to biomaterials: concept of biocompatibility in advances in biomaterials science and biomedical applications. INTECH 6:207–240
55. Armitage DA, Parker TL, Grant DM (2003) Biocompatibility and hemocompatibility of surface-modified NiTi alloys. J Biomed Mater Res A 66:129–137
56. Canfarotta F, Waters A, Sadler R, Mcgill P, Guerreiro A, Papkovsky D, Piletsky S (2016) Biocompatibility and internalization of molecularly imprinted nanoparticles. Nano Res 9:3463–3477
57. Bernier M, Paul RK, Martin-Montalvo A, Scheibye-Knudsen M, Song S, He HJ, Armour SM, Hubbard BP, Bohr VA, Wang L, Zong Y, Sinclair DA, De Cabo R (2011) Negative regulation of STAT3 protein-mediated cellular respiration by SIRT1 protein. J Biol Chem 286:19270–19279
58. Sindrilaru A, Peters T, Wieschalka S, Baican C, Baican A, Peter H, Hainzl A, Schatz S, Qi Y, Schlecht A, Weiss JM, Wlaschek M, Sunderkötter C, Scharffetter-Kochanek K (2011) An unrestrained proinflammatory M1 macrophage population induced by iron impairs wound healing in humans and mice. J Clin Invest 121:985–997
59. Hashemi-Moghaddam H, Zavareh S, Karimpour S, Madanchi H (2017) Evaluation of molecularly imprinted polymer based on HER2 epitope for targeted drug delivery in ovarian cancer mouse model. React Funct Polym 121:82–90

Chapter 8
MIP as Drug Delivery Systems of Ophthalmic Drugs

Long Zhao and Zhaosheng Liu

8.1 Sustained Release of Ophthalmic Drugs in Contact Lenses with Different Release Durations

8.1.1 Introduce

The human eye is composed of several barriers and defense mechanisms to protect itself from the surrounding environment. Because of this special anatomical structure, it is a great challenge to deliver drugs and treat ocular diseases. When the disease occurs in the eye (Fig. 8.1), systemic administration is often not effective or even ineffective because of the presence of the blood-ocular barrier. The blood-ocular barrier is composed of the blood-aqueous barrier (BAB) and the blood-retinal barrier (BRB). The BAB is located in the anterior segment of the eye, which is formed by the endothelial cells of the blood vessels in the iris and the non-pigmented cell layer of the ciliary epithelium [1]. There are tight junction complexes in both cell layers, which prevent nonspecific passage of solutes from entering the intraocular milieu, otherwise they may negatively affect the transparency and chemical balance of the ocular fluids [2].

The BRB [1] is located in the posterior part of the eye, which consists of endothelial cells of retinal blood vessels (inner blood-retinal barrier) and the retinal pigment epithelial cells (outer blood-retinal barrier). The role of the BRB is to regulate the penetration of substances from the blood into the retina, thus limiting drug delivery to retinal tissues and vitreous by systemic administration. BAB and BRB restrict the delivery of drugs from the systemic circulation to the anterior and posterior segments. After systemic administration, the intravitreal drug concentration is usually only 1–2% of the plasma concentration [3]. Therefore, ophthalmic

L. Zhao · Z. Liu (✉)
College of Pharmacy, Tianjin Medical University, Tianjin, China

Z. Liu et al. (eds.), *Molecularly Imprinted Polymers as Advanced Drug Delivery Systems*, https://doi.org/10.1007/978-981-16-0227-6_8

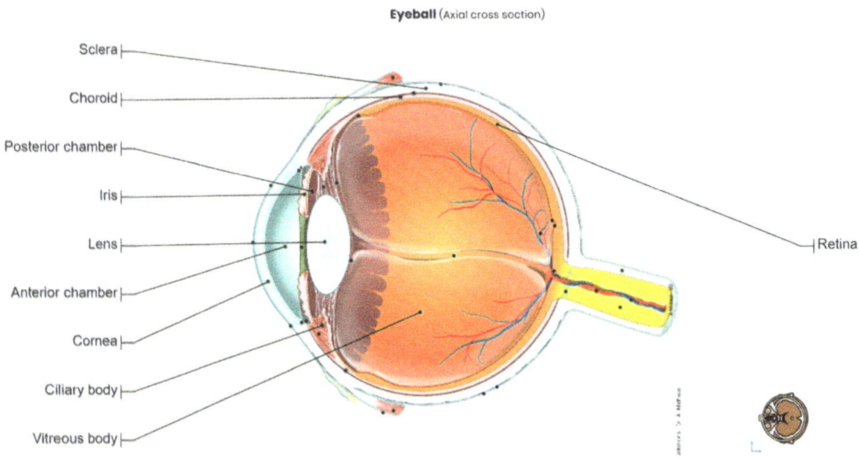

Fig. 8.1 The structure of the eye. The anterior segment: the cornea, conjunctiva, iris, ciliary body, lens, aqueous humor; the posterior segment: sclera, choroid, retina, and vitreous

drug delivery by the topical route is considered to be the most appropriate treatment route for ophthalmic diseases [4].

Common ophthalmic preparations include eye drops, suspensions, and ointments. Eye drops are still the most convenient and common method of ocular administration. It is well known that most drugs delivered in this way have difficulty in reaching aqueous humor. Predictions indicate that the proportion of lipophilic molecules is less than 5% and the proportion of hydrophilic molecules is less than 0.5% [5], which may be due to the tear drainage and removal, as well as the presence of the corneal barrier. When eye drops are not completely absorbed into the eyes, excessive amounts of drug can sometimes lead to unnecessary systemic bioavailability. In addition, the concentration of active ingredients of these drugs is usually very high. Therefore, despite the correct use of the recommended dose, a considerable amount of the drug may be absorbed in unnecessary ways through various channels, which may increase the possibility of side effects. At the same time, patient compliance also affects the success of administration.

Drug-laden CLs are developed on the basis of commercial CLs, which have been used for more than 50 years. Hydrogels are the main component of CLs, because their high water content and good performance make them highly compatible with human tissues [6]. The continuous development of a number of materials allows CLs to be worn for a long time. At present, more than 100 million of people around the world wear CLs to correct vision. If they are combined with drug delivery, the use of CLs will greatly increase, and the application prospects of CLs will be very broad [7].

Although general commercial CLs are very superior in light transmittance, water content, and comfort, the controlled and sustained release of drugs generally cannot be achieved if they are directly loaded with drugs. Burst release problems are common. For example, commercial lenses release ketotifen for a time ranging from 5 to 240 min [8]. Therefore, some techniques are used in the synthesis of CLs to transform them into the drug-loaded CLs with affinity for drugs or controlled release capabilities. These techniques or methods include: diffusion barrier created by Vitamin E [9], ion interactions [10], combination with cyclodextrin [11], supercritical solvent impregnation [12], incorporation of colloidal nanoparticles [13], molecular imprinting technology [14], and so on. One of the most important challenges of ocular administration is to maintain the optimal drug concentration at the site of action over a long period of time [15, 16].

Molecular imprinting technology (MIT) is trying to bypass these challenges by increasing the residence time of the drug to achieve controlled release. This technology uses drug molecules as templates in the polymerization process to induce the arrangement of monomers based on their affinity for drugs. The position of monomers is fixed. Once templates are removed, the imprinted cavities (imprinted sites) that interacts with drugs can be obtained. Compared with polymers synthesized without templates, molecularly imprinted polymers have higher affinity and interaction with drugs. This technology, which is initially developed for endowing rigid highly cross-linked polymeric systems with the ability to recognize target species, has been successfully adapted to the synthesis of CLs.

8.1.2 Antibacterials/Antibiotic

Norfloxacin (NRF) is one of the 4-quinolone antibacterial agents, which is believed to work by inhibiting the A subunit of the essential enzyme DNA gyrase in bacterial cells [17]. Carmen et al. manufactured MIP-CLs with norfloxacin [18], which had a high load on norfloxacin. Structurally, norfloxacin molecule has two ionizing groups: carboxylic acid ($pK_{a1} = 6.34 \pm 0.06$) and amino ($pK_{a2} = 8.75 \pm 0.07$), which can interact with the ionizing groups of other molecules. In addition, it can also establish hydrophobic interactions through hydrogen bond interactions or through aromatic rings.

Acrylic acid (AA) is a weak acid ($pK_a = 4.5$), that could interacts with protonable amino groups or hydrogen bond acceptor groups, thus was chosen as the functional monomer. The imprinted lens loaded more NRF than the non-imprinted polymer-based contact lenses (NIP-CLs), and the imprinting factor (IF, defined as the ratio of the drug load of MIP-CLs and NIP-CLs) can reach 2.46. They studied the effects of different NRF/AA ratios on release. When NRF/AA ratio was 1:3 and 1:4, respectively, the lenses had the strongest ability to control the release process and the time can be maintained for more than 24 h.

Ciprofloxacin is a second-generation fluoroquinolone drug. It interferes with bacterial DNA gyrase and prevents bacterial DNA replication [19]. It is a broad-spectrum antibiotic that inhibits both Gram-negative and Gram-positive bacteria [20, 21].

Acetic and acrylic acid were used as the functional monomers [22], to interact with the ciprofloxacin template to efficiently create recognition cavities within the final MIP. The synthesized materials were loaded with 9.06 , 0.10, and 0.025 mM solutions of ciprofloxacin, and the release of ciprofloxacin into an artificial tear solution was monitored over time. The materials can release drugs for periods varying from 3 to 14 days, dependent on the loading solution, functional monomer concentration, and functional monomer/template ratio. The materials made with greater monomer/template ratio (8:1 and 16:1, mol/mol) tended to release for longer periods of time. The materials with a lower monomer/template ratio (4:1) tended to release comparatively greater amounts of ciprofloxacin into solution, but the release was somewhat shorter. This work is the first to demonstrate the feasibility of molecular imprinting in model silicone hydrogel-type materials.

8.1.3 Antifungals

Ocular fungal infections are increasingly recognized as an important cause of morbidity and blindness; certain types of ocular fungal diseases may even be life-threatening [23]. Keratitis (corneal infection) is the most common manifestation, and may also affect the conjunctiva, sclera, eyelids, lacrimal organs, and intraocular structures [24]. In 2008, Byrne published a patent titled with therapeutic CLs with antifungal delivery (WO2008060574). The system used the principle of molecular imprinting (or called the principle of bionics) to select biological templates and functional monomers, which imitated the recognition of drugs and receptors, and initiated reactions to synthesize hydrogel CLs.

Macromer units or oligomer units were co-polymerized into the hydrogel matrix to form receptor locations within the hydrogel matrix that resembled or mimicked the receptor sites or molecules associated with the biological target tissue to be treated with the drug or the biological mechanism of action. In this patent, it shows fluconazole binding isotherm and dynamic release in artificial lacrimal solution of 5% cross-linked poly-HEMA-*co*-AM-*co*-DEAEM-*co*-NVP-*co*-PEG200DMA hydrogel lenses at room temperature 25 °C. It also pointed out that the ophthalmic drug delivery system and method were used to deliver antimicrobial agents. This method was also suitable for other different kinds of antimicrobial or antifungal agents.

8.1.4 Dry Eye Disease

Dry eye syndrome (DTS) is one of the most common conditions in ophthalmology and is often ignored. The 2017 report of the Tear Film and Ocular Surface Society International Dry Eye Workshop II (TFOS DEWS II) defines dry eye as "a multi-factorial disease of the ocular surface characterized by a loss of homeostasis of the tear film, and accompanied by ocular symptoms, in which tear-film instability and hyperosmolarity, ocular surface inflammation and damage, and neurosensory

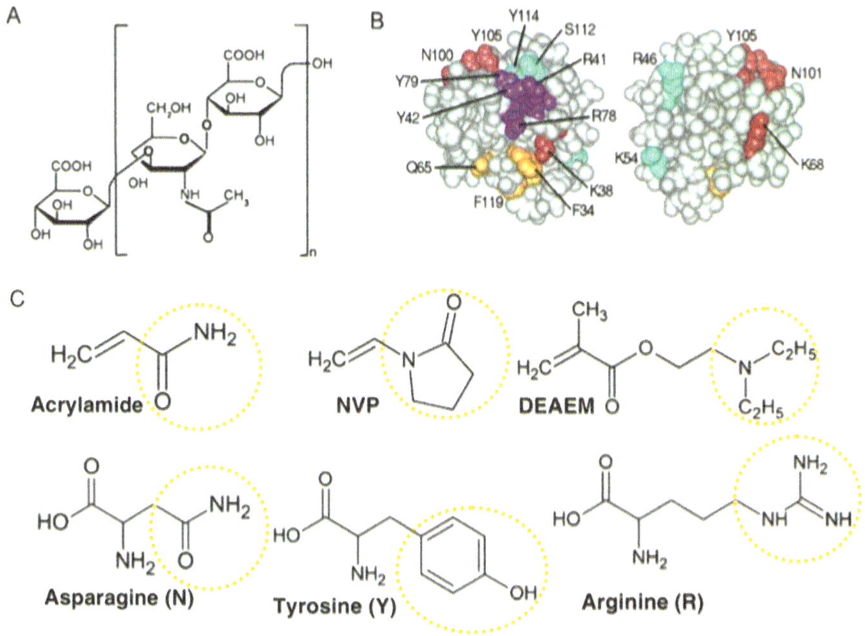

Fig. 8.2 The structure of hyaluronic acid and biomimetic approach: comparison of functional groups on amino acids and monomers. (**a**) Hyaluronic acid is a long-chain molecule, specifically a polysaccharide consisting of repeating units of glucuronic acid and *N*-acetylglucosamine (dimer molecular weight is 415 Da). (**b**) Hyaluronic acid binding protein CD44, is a naturally found protein with a high affinity for hyaluronic acid. (**c**) For the biomimetic imprinting of hyaluronic acid, acrylate and methacrylate monomers were selected that bear chemical similarity to the amino acids found on the binding site CD44. Acrylamide and asparagine both have amide moieties, *N*-vinyl pyrrolidone and tyrosine have hydrogen bonding capability while 2-(diethylamino)ethyl methacrylate is positively charged, like arginine and lysine. (Reproduced with permission of [29])

abnormalities play etiological roles" [25]. As a common ocular surface disease, dry eye affects over 20 million people worldwide, with varying degrees of severity, and this number will increase [26].

Hyaluronic acid (HA) is a linear glycosaminoglycan made up of repeating units of *N*-acetyl-D-glucosamine [27]. It is a topical artificial tear, and has a good effect in treating eye discomfort and dry eye syndrome [28]. Using molecular imprinting technology, Maryam reasonably designed and synthesized a hydrogel contact lens that could control the release of HA [29]. By analyzing the structure of the HA receptor, CD44, the suitable functional monomers were selected according to the structural characteristics of amino acid residues (Fig. 8.2). During the synthesis of the MIP-CLs, the modified polyvinyl alcohol (PVA) monomers were used as functional monomers, which were synthetic raw material for commercial lenses nelfilcon A [30].

Delayed release characteristics were significantly improved through biomimetic imprinting, as multiple functional monomers provided non-covalent complexation points within nelfilcon A gels without altering structural, mechanical, or optical properties. As the concentration of functional monomers decreased to 0.25% or 0.125%, the HA released gradually increased. As the proportion of diethylaminoethyl methacrylate (DEAEM) increased, the cumulative release quality showed a downward trend. Therefore, the release of HA can be adjusted according to the amount and proportion of functional monomers in the network.

At the same time, the factors affecting the diffusion coefficient of the drug were also studied and analyzed. The diffusion coefficient of HA was controlled by varying the number and variety of functional monomers (increasing the variety of functional monomers lowered the HA diffusion coefficient 1.5 times more than single functional monomers, and 1.6 times more than nelfilcon A alone). This is the first demonstration of imprinting a large molecular weight polymer within a hydrogel.

In addition to hyaluronic acid, hydroxypropyl methylcellulose (HPMC) often acts as the main component of artificial tear eye drops. It is used to relieve the discomfort of eyes such as dry eyes or burning and tingling caused by exposure to the sun or wind and sand. It is also a protective agent to prevent further irritation. There are some other cellulose derivatives, such as carboxymethyl cellulose (CMC) [31].

One MIP-CLs of HPMC were announced in 2011 [32]. The authors used a material named Lotrafilcon B, which was got from CIBA Vision, Inc. This material is made from betacon macromer, methacryloxypropyl-tris-(trimethylsiloxy) silane (Tris) and dimethyl acrylamide (DMA). The MIP-CLs were synthesized using functional monomer AA, imprinted molecule HPMC, photo-initiator Darocur 1173 and cross-linking agent polyethylene glycol (200) dimethacrylate (PEG200DMA), as well as ethylene glycol dimethacrylate (EGDMA). Better light transmittance and mechanical properties can be achieved. In the in vitro release experiment, both the release conditions of the perfect sink model and a flowing microfluidic device were used. When thin lenses were in a perfect sink environment, the release of HPMC was completed in 10–12 days. The release of HPMC was at a linear rate of 16 μg HPMC/day for 60 days, in an ocular flowrate (3 mL/min) provided by the flowing microfluidic device. This work highlights the potential of imprinting in the design and engineering of silicone hydrogel lenses.

8.1.5 Glaucoma

Glaucoma is a group of irreversible progressive optic nerve diseases, characterized by optic neuropathy and visual-field damage, which can lead to severe visual-field loss and blindness. In 2010, among the 32.4 million blind people in the world, glaucoma was the cause of blindness in 2.1 million people (6.5%) [33]. The only proven and generally accepted treatment to reduce the risk of further development of glaucoma optic neuropathy is to lower intraocular pressure and prevent the further development of glaucoma optic nerve damage [34, 35]. The commonly used drugs for lowering intraocular pressure are as follows: (a) prostaglandin analogs; (b) β-adrenergic blockers; (c) carbonic anhydrase inhibitors; and (d) miotics.

8.1.5.1 Bimatoprost

Prostaglandin analogs include bimatoprost, latanoprost, tafluprost, travoprost, and unoprostone. Bimatoprost is widely used in the treatment of glaucoma. Recently, the contact lens against bimatoprost [36] was fabricated with EGDMA, dimethyl acrylamide, siloxane, AA, and HEMA. The reaction can be completed at UV 370–380 nm for 15 min. Compared with the drug-free control contact lens, the optical transmittance and swelling degree of the MIP drug-loaded lenses did not change. The MIP-CLs showed that the ability to take drugs from the packaging solution increased by 125.25–137.20%. Due to the improved release of the drug by molecular imprinting, bimatoprost can be released continuously for 36–60 h. In contrast, the release time of the NIP-CLs was 24–36 h. The average release rate of the MIP-CLs within 60 h was 66.7 ng/h, which was more than 58.3 ng/h of the NIP-CLs.

8.1.5.2 Timolol

β-Adrenergic blockers include betaxolol, carteolol, levobunolol, metoprolol and timolol. As early as 2002, the first work of the MIP lens against β-adrenergic blockers was reported [37] using HEMA as the main chain monomer, MAA or methyl methacrylate (MMA) as the functional monomer, EGDMA as the cross-linking monomer, and timolol maleate as the imprinted molecule (Fig. 8.3). The release rate of timolol depended on the nature and quantity of the co-monomer in the medium and hydrogel. When the pH > 7, because most of the MAA groups were ionized, as the hydrophilicity of the hydrogel increased, and the ability to interact with the drug through hydrogen bonds or hydrophobic bonds decreased, thereby increasing the release rate. It was also observed that in the drug solution of pH 5.5, the imprinted lens prepared with the lowest MAA ratio was loaded with timolol three times that of the NIP-CLs.

Further improvement of the loading capacity of weakly cross-linked hydrogels can be obtained after replacing HEMA with *N,N*-Diethylacrylamide (DEAA) [38]. When EGDMA was 60–280 mM, the maximum adsorption capacity of the MIP-CLs was 2.26–2.68 mM, while that of the NIP-CLs was only 0.08–0.51 mM. The minimum cross-linker concentration for the imprinting to be effective was 80 mM EGDMA. Moreover, the release of the MIP-CLs in 0.9% NaCl exceeded 24 h.

Subsequently, the effect of different backbone monomers on the imprinted soft CLs of the timolol delivery system was studied [39]. The effect of four backbone monomers or their combination was studied, i.e., DEAA, HEMA, MMA/DMAA (50:50 v/v) and 1-(tristrimethyl-siloxysilylpropyl)-methacrylate (SiMA)/*N,N*-dimethylacrylamide (DMAA) (50:50 v/v). The total affinity of the imprinted soft CLs to timolol was in the order of HEMA > SiMA/DMAA > MMA/DMAA > DEAA. In 0.9% NaCl solution, the MMA/DMAA and SiMA/DMAA

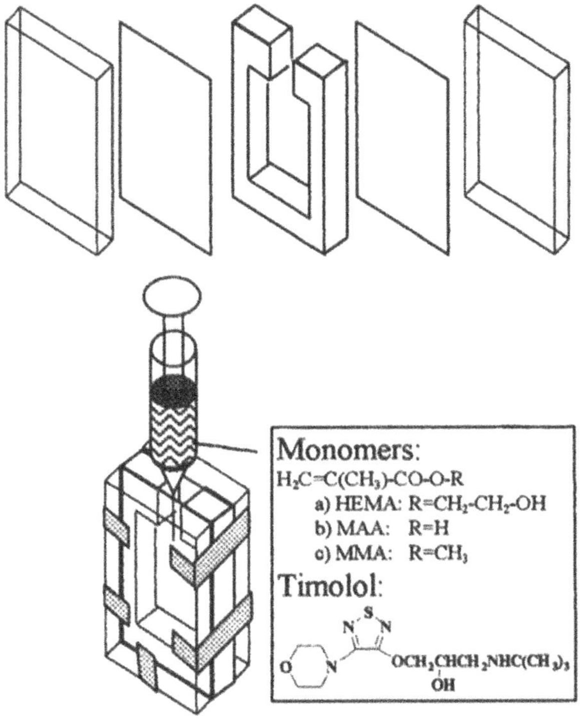

Fig. 8.3 Assembling and filling the mold before polymerization. (Reproduced with permission of [37])

lenses released the full dose in less than 3 h, while the HEMA and DEAA lenses showed a sustained-release curve over 9 h. The results obtained indicate that it is possible to adapt the drug loading and release behavior of the lenses to the treatment requirements of specific pathological processes by modulating the composition of the lenses.

8.1.5.3 Acetazolamide and Ethoxazolamide

Carbonic anhydrase inhibitors (CAIs) are also a class of drugs that reduce intraocular pressure to treat glaucoma. Common drugs are acetazolamide (ACT) and ethoxazolamide (ETOX). Andreza et al. first analyzed the structure of carbonic anhydrase (CA), which was the binding site of CAIs [40]. The active center of CA is a cone-shaped cavity where zinc ion is the center. The zinc ion coordinates with three histidine residues (His96\His94\His119), and there is the fourth ligand (Thr199). Inhibition can be achieved using CAIs by combining with zinc ions.

For that reason, they chose monomers whose chemical groups were similar to the amino acids at the active binding site. The zinc ion for CA was provided by methacrylates ($ZnMA_2$) represented, hydroxyl and amino groups were provided by HEMA and *N*-hydroxyethyl acrylamide (HEAA) and histidine was provided by

4-vinylmidazole (4VI) or 1-vinylmidazole (1VI). EGDMA was the cross-linking agent, and ACT was the imprinted molecule. The NVP-*co*-DMA hydrogels bearing 4VI, HEAA and Zn^{2+} showed twofold increase in drug affinity (estimated as network/water partition coefficient) and more sustained delivery. The ACT-imprinted networks gave the highest loading and controlled ACT release for 9 h. ETOX release was sustained for more than 1 week. It had favorable physicochemical and mechanical and cytocompatibility features.

8.1.5.4 Dorzolamide

Dorzolamide (DZD) is a kind of CAIs. The drug developed by Merck was introduced to the market in 1995. It is generally used in the form of hydrochloride to treat glaucoma. In the work of Malaekeh et al. [41], they used HEMA as the backbone monomer, MAA as the functional monomer, dorzolamide as the imprinted molecule, and EGDMA was used as a cross-linker monomer to synthesize the MIP-CLs and NIP-CLs. When the DZD/MAA molar ratio was 1:4, the MIP-CLs had the highest drug loading capacity, and the strongest ability to control the release process in an aqueous medium.

8.1.6 Inflammation

8.1.6.1 Prednisolone Acetate

Prednisolone acetate (PA) is a glucocorticoid with a strong anti-inflammatory effect, and used to reduce non-infectious inflammation. Bizhan et al. published a related study in 2011 [42]. HEMA was the backbone monomer. EGDMA was a cross-linker. MAA was a functional monomer (0, 50, 100, and 200 mM, respectively). AIBN was as the initiator.

The imprinted lens whose PA/MAA ratio was 1/4 was the best in all aspects. The water content reached 62%, and there was no difference between the water content of the MIP-CLs and NIP-CLs. The molecular imprinting technology had no effect on the swelling degree and water content of lenses. The drug loading of the MIP-CLs was 58 μg/disc, while the drug loading of the NIP-CLs was 39 μg/disc. In artificial tears, the MIP-CLs released about 64% of PA within 48 h, while the NIP-CLs released 78% within 8 h. These results provide the possibility to prepare hydrogel soft CLs as a drug delivery system of glucocorticoid.

8.1.6.2 Diclofenac Sodium

In addition to glucocorticoids, non-steroidal anti-inflammatory agents (NSAID) is also a class of drugs used to treat inflammation. Diclofenac sodium is a small molecule drug that dissociates to more than 99% at physiological pH (7.4)

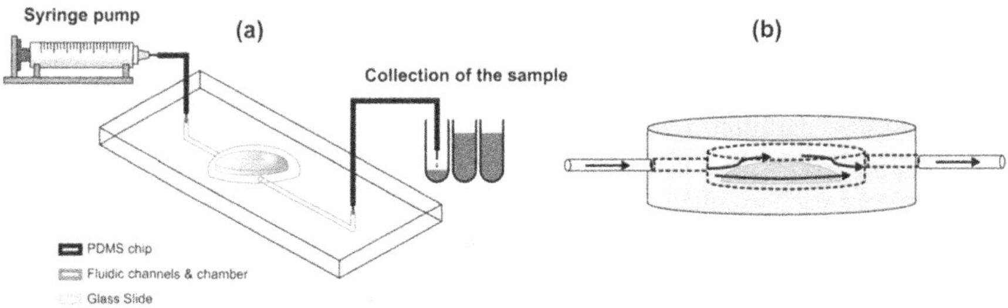

Fig. 8.4 Microfluidic chip design with physiological ocular flow. (**a**) Schematic of experimental set-up for contact lens drug delivery evaluation. (**b**) The inner chamber has a radius of curvature of 9.00 ± 0.10 mm. A drug-loaded lens is placed over a mount with radius of curvature of 8.75 ± 0.10 mm and the device is sealed against a glass plate. (Reproduced with permission of [44])

[43]. Arianna [44] chose a functional monomer diethylaminoethyl methacrylate (DEAEM) ($pK_a = 9.2$). The interaction between DEAEM and the drug is mainly ionic force. Ionic bond is the strongest non-covalent bond, which is close to one third of the strength of covalent bond and 20–30 times that of hydrogen bond [45].

The lenses were composed of 90 mol% backbone monomer HEMA, 5 mol% functional monomer DEAEM and 5 mol% cross-linking monomer PEG200DMA. When the *M/T* ratio were 10.5, 3.5 and 1.0, the maximum loading capacity of the imprinted hydrogel was increased by 67%, 76%, and 83% respectively, compared with the NIP-CLs. Moreover, the loading of diclofenac of the MIP-CLs was increased by about five times, compared with the MIP-CLs without DEAEM. This study used a new fluid device to perform in vitro release experiments (Fig. 8.4). The device was designed to simulate the flow rate of tears. In the first 24 h, the lens with an *M/T* ratio of 10.5 released diclofenac at a constant rate of 8.62 g/h, which was close to the maximum dose of commercial eye drops, as a result, the MIP-CLs had great application potential.

8.1.7 Allergy

In the past few decades, the incidence of eye allergy has increased all over the world [46]. The most common eye allergy is caused by allergic conjunctivitis (SAC). Drugs that combine powerful antihistamine effects and stable mast cell properties have become SAC's first-line drugs. The most common representatives are ketotifen, azelastine, and olopatadine.

8.1.7.1 Olopatadine

Olopatadine is considered a multimodal anti-allergic drug. It also shows anti-inflammatory properties. It is a selective histamine H1 receptor antagonist [47]. It is the only one that can stabilize hypertrophy with the concentration of commercially available eye drops cellular drugs [46, 48]. Clara et al. chose the functional monomers similar to the chemical groups of amino acids in the active site of the H1 receptor [49]. The backbone monomer was HEMA, the cross-linking monomer was EGDMA, and the functional monomer was a combination of four monomers, including AA, AM, 2-Acrylamido-2-methyl-1-propanesulfonic acid (AMPSA) and benzylmethacrylate (BzMA). AIBN was the initiator.

In 400–700 nm, the transmittance of all lenses was above 97%. The swelling degree of most lenses was similar to that of traditional lenses based on p-HEMA (over 80%). The optimized hydrogels provided in few hours olopatadine concentrations was similar to those of commercially available eye drops but the levels could be sustained for a whole day, demonstrating their efficacy. The olopatadine-loaded CLs successfully passed the HET-CAM test of ocular irritancy and showed good compatibility with mast cells. They were able to inhibit the release of histamine and TNF-α from sensitized mast cells, proving their potential application in preventing and treating allergic conjunctivitis.

8.1.7.2 Ketotifen

Ketotifen fumarate is a powerful molecule with high selectivity for histamine H1 receptor [50]. Mark E. Byrne's team analyzed the structure of the H1 receptor and searched for functional monomers with similar structures through key amino acid residues [14]. Eventually, they chose AA, AM, and *N*-vinyl 2-pyrrolidinone (NVP) as functional monomers. Lenses were synthesized by 3% functional monomer, 92% backbone monomer HEMA and 5% cross-linking monomer PEG200DMA.

The combination of functional monomers was investigated, and the best result obtained was the combination of three functional monomers, which increased the drug loading of the MIP-CLs by six times compared with the NIP-CLs (Fig. 8.5). In addition, the mechanical and optical properties of the lenses were found to be consistent with traditional lenses. In artificial tears, the MIP-CLs could be released continuously for 2 days. However, when lysozyme (1 mg/mL) was added to artificial tears, the MIP-CLs could last for 5 days of extended release, in which 80% of the drug was released in about 4 days.

Mark E. Byrne also studied the release dynamic drug in vitro in a new type of microfluidic device (Fig. 8.6) [51]. It only simulated the volume flow rate of the eye, tears volume and tear composition. The lens (AA-*co*-AM-*co*-NVP) exhibited a ketotifen fumarate diffusion coefficient of 5.57×10^{-10} cm^2/s, which was a factor of 9, 7.2, and 13.8 less than the AA-based lens, AM-based lens, and AA-co-AM-based lens, respectively. For the poly(AA-*co*-AM-*co*-NVP-*co*-HEMA-PEG200DMA) lens, the drug was slowly released at a constant zero-order release rate for about 3.5 days, and the total release time was more than 1 week. Taking into account that the size of the lens was significantly different from that of commercial lenses, the release amount of the lens was 0.8–1.4 μg/day. The estimated dose of eye

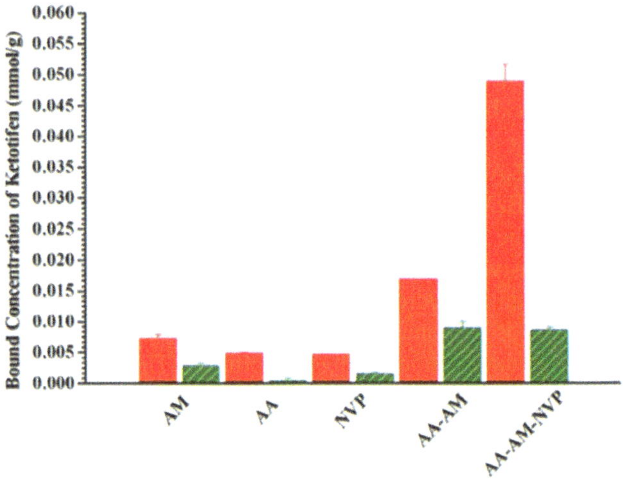

Fig. 8.5 Enhanced loading of ketotifen for multiple monomer gels for poly(*n*-*co*-HEMA-*co*-PEG200DMA) networks at 0.4 mg/mL loading concentration. (Reproduced with permission of [14])

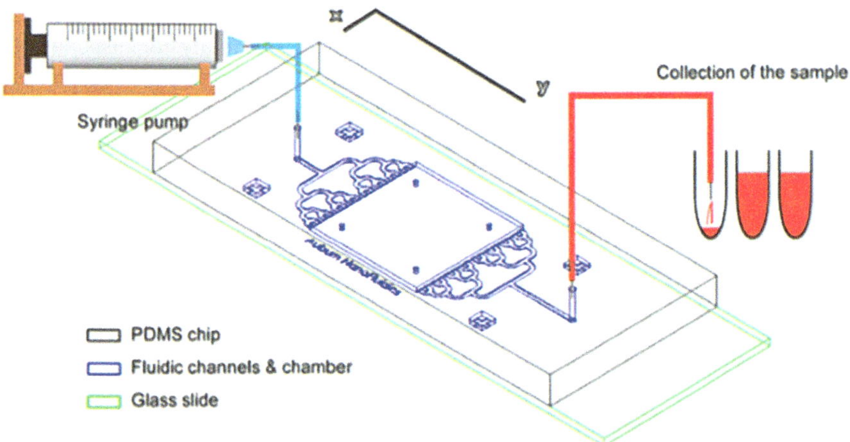

Fig. 8.6 Microfluidic chip design with physiological ocular flow. Schematic of experimental set-up for contact lens drug delivery evaluation. The hydrogel is placed in the microfluidic chamber (chamber height—560 μm, and width—1600 μm) between the four posts and drug release is measured within artificial lacrimal fluid flow rates. (Reproduced with permission of [51])

drops was 1–1.5 μg/day. It demonstrates the enormous potential for molecular imprinting to further tailor therapeutic release kinetics via the imprinting process.

8.2 The Entrapment of Biological Ophthalmic Drugs by MIP

The purpose of molecular imprinting technology is to optimize the affinity for drugs through the spatial arrangement of functional monomers. However, the preparation of hydrogels for peptide drugs is still challenging. There are two main reasons: (a) the peptide must be dissolved in the monomer solution and be able to diffuse in the network during removal and rebinding, which is difficult due to the steric hindrance of the network; (b) the cross-linking density of CLs is relatively low, and their swelling in aqueous media (including tear fluid) after polymerization may reduce the stability of imprinted cavities [52]. Therefore, the design of the MIP DDS should produce the best network stability to ensure maximum interaction between the network and the drug molecules. Relative to small molecules, peptide drugs may not be as effective as small molecules due to their complex structure. However, due to the existence of functional monomers and the many active sites on the molecular structure of peptide drugs, it is possible to accomplish good imprinting by careful selection of composition.

Polymyxin B is a group of polypeptide antibiotics produced by bacillus polymyxa. It exhibits rapid activity against multidrug-resistant Gram-negative bacteria [53, 54]. A hydrogel contact lens suitable for loading and releasing polymyxin B has been developed [55]. The NIP-CLs A and MIP-CLs B–G were synthesized according to the scheme in the Table 8.1.

The oxygen permeability (D_k) of the lenses was about 65–80 × 10^{-11} cm^3 cm^2/(cm^3 s mmHg), the water content was about 50%. Lenses exhibited a relatively low transmittance. In the range of 300–600 nm, the maximum transmittance of all lenses was about 45%.

In terms of drug loading capability, the ratio of AA, EGDMA, drug and water, all played an important role in the loading performance of polymyxin B of lense. The imprinted lens significantly increased the loading of polymyxin B. It was also interesting that a small amount of water added to the polymerization system could significantly promote the loading capability. The water was conducive to the dissolution of the drug before polymerization and the interaction with AA, and it could also increase the mesh size of the hydrogel. Moreover, when the ratio of

Table 8.1 Hydrogels composition. (Reproduced with permission of [55])

	Hydrogel						
Components	A	B	C	D	E	F	G
Polymyxin B (mg)	0	12.5	12.5	12.5	25	50	50
HEMA (mL)	4.5	4.5	4.5	4.5	4.5	4.5	4.5
AA (mL)	0.2	0.2	0	0.2	0.2	0.2	0.2
Water (mL)	0	0	0	0	0	0.2	0
EGDMA (mL)	0.3	0.3	0.3	0.15	0.3	0.3	0.3
AIBN (mg)	10	10	10	10	10	10	10

cross-linking agent was low, the loading capacity of lens was the highest, which indicated that an increase in mesh size was beneficial to the entry and interaction of the antimicrobial peptide.

Acrylic acid-functionalized and imprinted hydrogels loaded greater amounts of polymyxin B and led to more sustained-release profiles. The lens could be used for the hosting of other related antimicrobial peptides (for instance vancomycin). In the microbiological tests, lenses all showed the ability to inhibit bacterial infections. In the ICCVAM-recommended hen's egg test-chorioallantoic membrane (HET-CAM) tests, none of the polymyxin B-loaded lens caused hemorrhage, lysis or coagulation in the chorioallantoic membranes, which suggested adequate biocompatibility in spite of their potent antibacterial effect.

8.3 In Vivo Studies of Ophthalmic Drug-Imprinted Polymers

Most MIP-CLs still remains utility in in vitro evaluation to investigate the effects of modifications or synthesis procedures on drug release kinetics and to identify candidate material/drug combinations for in vivo testing. However, it is only through testing in animal and human models that the potential advantages of these systems can be identified. Thus far, in vivo testing of such systems suggests that the field is moving in the right direction to reach these design goals, as in vivo evidence that contacts lens drug delivery systems improving ocular residence times and treatment outcomes has been published.

8.3.1 Ciprofloxacin

Alex et al. synthesized ciprofloxacin sustained-release silicone hydrogel CLs in 2014 and studied the effects of lenses in the treatment of microbial keratitis (MK) in vivo and in vitro [56]. In the in vivo experiment, they used rabbit scratch model of MK. Then the cornea was removed for counting colony-forming units (CFU). The cornea treated with ciprofloxacin eye drops completely sterilized the cornea within a short period of 8 h of treatment. The therapeutic effect of the control lens (or named NIP-CLs) was not significant, and the CFU was almost the same as the non-intervention group. The MIP-CLs with monomer/template ratio of 4/1 and 8/1 achieved a certain antibacterial effect, although they did not completely kill all the bacteria.

8.3.2 Bimatoprost

In the anti-glaucoma section of Sect. 8.1.5.1, we described the synthesis and in vitro release of bimatoprost. Here we discuss the in vivo study [36]. In vivo rabbit tear data showed that the retention time of bimatoprost in the MIP-CLs was improved compared with non-imprinted glasses and eye drops.

The experiment was performed using New Zealand rabbits (male and female) to investigate the retention time of bimatoprost in rabbit tears. There were three experimental groups: (a) bimatoprost-loaded lenses (SM-20, 16.15 ± 3.34 μg); (b) bimatoprost-loaded MIP-CLs (MP-10, 14.76 ± 3.22 μg) and (c) 0.03% bimatoprost eye drops (1 drop ≈ 50 μL ≈ 15 μg). The lens was implanted into the right eye of the rabbit without local anesthesia, and the left eye was used as a control. The total amount of the three groups was approximately equal. The C_{max} (5 min) of eye drops, the NIP-CLs and imprinted lens were 145.26, 62.35 and 56.26 μg/mL, respectively. The concentration of the eye drops dropped rapidly within 1 h and there was no drug detected after 1 h. The NIP-CLs and imprinted lens were slowly released within 12 h. The drug concentration of imprinted lens was always greater than that of the NIP-CLs. The drug concentration of both at 12 h was close to the concentration of eye drops at 1 h (Fig. 8.7). It illustrates the successful application of molecular imprinting strategy in the drug-loaded lens of bimatoprost.

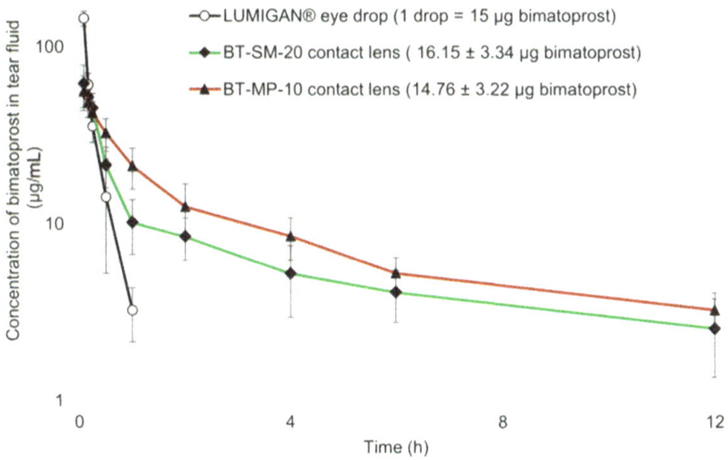

Fig. 8.7 Bimatoprost tear fluid concentration using CLs and eye drop solution ($n = 6$). (Reproduced with permission of [36])

8.3.3 Timlol

The *in vivo* study of timolol MIP-CLs was the continuation of the second work of Haruyuki Hiratani and Carmen Alvarez-Lorenzo's research group [57] (See Sect. 8.1 for the lens composition scheme). Male Nippon albino rabbits weighing 3.3–3.7 kg were chosen as experimental animals. The lenses were soaked in 10 mL timolol solution (1 mM) for 3 days to prepare drug-loaded lenses, rinsed with water, and then sterilized (20 min, 121 °C, 0.2 MPa). There were four experimental groups: (a) MIP-CLs with a drug loading of 34 μg; (b) NIP-CLs with a drug loading of 21 μg; (c) administer 34 μg by 0.068% timolol eye drops; (d) administer 125 μg by 0.25% timolol eye drops. After collecting the tear fluid sample with a glass capillary, it was diluted and measured.

In terms of the release time, the imprinted lens released 180 min, which was twice of the NIP-CLs (90 min). In contrast, when 0.068% and 0.25% of eye drops for timolol was used, the drug disappeared from the tears within 60 min. C_{max} for the imprinted lens was 330.88 μM, which was 1.5 times that of the NIP-CLs, 3 times that of 0.068% eye drops and 1.4 times that of 0.25% eye drops. Moreover, the concentration–time curve (AUC) of the imprinted lens was the highest, which was significantly higher than that of the other groups.

8.3.4 Ketotife

Mark E. Byrne designed a better imprinted lens of ketotifen fumarate (anti-allergic part in Sect. 8.1), and carried out an in vivo study using the lens [58]. The in vivo experiment was divided into three intervention groups: (a) imprinted lens; (b) NIP-CLs; (c) a drop of 0.035% ketotifen eye drops (50 μL, 17.5 μg ketotifen fumarate). The drug loading of ketotifen fumarate in the poly (HEMA-*co*-AA-*co*-AM-*co*-NVP-*co*-PEG200DMA) imprinted lens was 115 ± 10 μg/lens, and the NIP-CLs was 39 ± 2 μg/lens. The lenses were placed on the right cornea of each rabbit. 3–5 μL of tear fluid was collected from the lower eyelid and conjunctival sac at regular intervals. The NIP-CLs quickly reached C_{max} (140.00 ± 35 μg/mL). The concentration decreased exponentially and gradually disappeared within 10 h (Fig. 8.8). The NIP-CLs maintained the effective dose (ED_{50}) (30 μg/mL) for only 7 h. The imprinted lens reached C_{max} (214 ± 63 μg/mL) at 4 h, and had a relatively constant concentration (170 ± 30 μg/mL) within 26 h.

For the eye drops, an exponential decline within 45 min was observed. This meant that most of the medicine was quickly excreted by the tear fluid. The time of ED_{50} was only 30 min. The AUC of the imprinted CLs was 4365 ± 1070 μg h/mL, which was almost 9 times that of the NIP-CLs and 94 times that of eye drops. Mean residence time (MRT) of ketotifen fumarate in the imprinted lens was

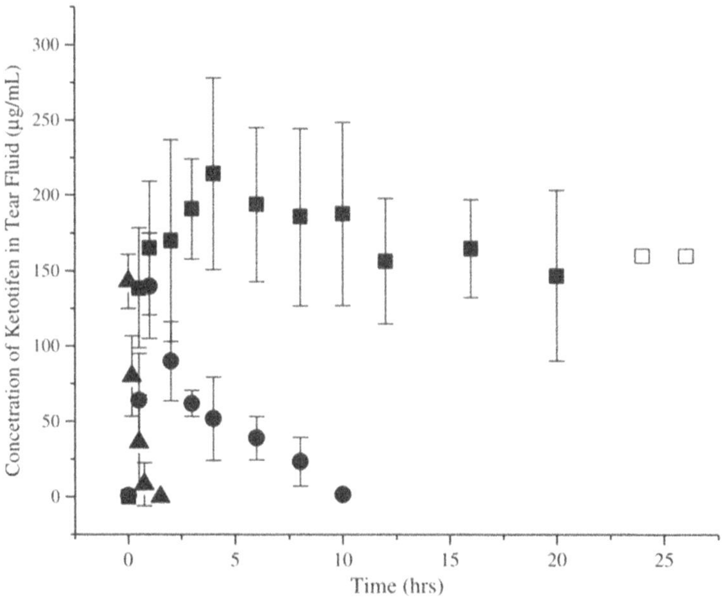

Fig. 8.8 In vivo ketotifen fumarate tear fluid concentration profile from CLs and topical eye drops in a white New Zealand rabbit. Dynamic ketotifen fumarate release from poly (HEMA-co-AA-co-AM-co-NVP-co-PEG200DMA) CLs, Imprinted (■, □) and non-imprinted (•). (Lens details: 100 ± 5 μm thickness, diameter 11.8 mm, no power, loaded at 0.3 mg/mL ketotifen concentration). Release from one eye drop (0.035% solution Zaditor® Novartis) (▲). Imprinted CLs clearly demonstrate an extended release of ketotifen fumarate, a significantly increased ketotifen fumarate residence time, and a significantly increased bioavailability of ketotifen fumarate in the tear fluid for an extended duration compared to non-imprinted lenses and topical drop therapy. Solid data points represent the mean ± SD (N = 3–5) and hollow data points represent single run (N = 1). (Reproduced with permission of [58])

12.60 ± 0.37 h, which was 3.75 times than that of the NIP-CLs and almost 50 times that of eye drops. All the information of the lenses mentioned were summarized in Table 8.2.

8.4 Conclusion

As a whole, molecular imprinting provides an exciting rational engineering strategy for sustained release. The imprinted lens can give a constant drug concentration over a longer period of time modulated by incorporating adequate functional monomers. It is clear that imprinted lenses are very promising combination devices and are much more effective and efficient delivery devices than eye drops. Additional studies on various therapeutic groups and different chemical structures are needed to extend the scope of application of the imprinted lenses as therapeutic devices.

Table 8.2 The summary of MIP-CLs for ophthalmic drugs

Category	Drug	Functional monomers	Backbone monomers	Crosslinker	Size (diameter/tickness)	Drug loading	Release medium	In vitro results	In vivo results	Note	Reference
Antibacterials	Norfloxacin	AA or NVP	HEMA	EGDMA	10 mm/0.9 mm, 0.4 mm	0.18 mg/lens	10–15 mL ALF	24 h		The ratio of template/functional monomer was studied by isothermal titration calorimetry (ITC)	[18]
	Ciprofloxacin	HOAc or AA	HEMA, TRIS	EGDMA	/0.9 mm	N/A	2 mL ALF	3–14 days		Low water content (15%)	[22]
		AA	HEMA, TRIS, PVP	EGDMA	/64 μm	1133–1509 μg/lens	2 mL PBS	3 days	Rabbit scratch model of microbial keratitis	Lenses had a certain antibacterial effect, not as good as eye drops. The antibacterial effect was evaluated by the number of colony-forming units	[56]
	Polymyxin B (peptide)	AA	HEMA	EGDMA	10 mm/0.4 mm	70–100 mg/g	2 mL 0.9% NaCl	14 days		Low transmittance (15–45%), HET-CAM tests showed good biocompatibility	[55]
Antifungals	Fluconazole	DEAEM, AM, NVP	HEMA	PEG200DMA	N/A	N/A	ALF	2 days		Patent number WO2008060574	

Dry eye disease	Hyaluronic acid	DEAEM, AM, NVP	Modified PVA		/127 μm	N/A	20 mL ALF	5 days		Functional monomers were selected by analyzing receptor structure, PVA was the material of nelfilcon A (CIBA Vision, Inc.)	[29]
	Hydroxypropyl methylcellulose	AA, DMA	TRIS, Betacon Macromer	EGDMA, PEG200DMA	/100 μm	1050 mg/lens	3 μL/min DI water	60 days (16 mg/day)		Betacon Macromer was the material of Lotrafilcon B (CIBA Vision, Inc.).	[32]
Glaucoma	Bimatoprost	DMA, siloxane, MAA	HEMA	EGDMA	N/A	14–28 μg/lens	2 mL ALF	60 h	12 h	Eye drops was for 1 h, drug concentration of lens at 12 h was close to that of eye drops at 1 h	[36]
	Timlol	MAA or MMA	HEMA	EGDMA	14 mm/ 0.7 mm	12 mg/g	15 mL 0.9% NaCl, PBS, ALF	12 h		Drug loading was affected by pH.	[37]
		MAA	DMA	EGDMA	/0.3 mm	2.26–2.68 mM	10 mL 0.9% NaCl	24 h			[38]
		MAA	HEMA or TRIS/DMA or MMA/DMA or DEAA	EGDMA	/0.3 mm	2.6–8 mM	10 mL 0.9% NaCl	9 h			[39]

(continued)

Table 8.2 (continued)

Category	Drug	Functional monomers	Backbone monomers	Crosslinker	Size (diameter/ tickness)	Drug loading	Release medium	In vitro results	In vivo results	Note	Reference
		MAA	DMA, TRIS	EGDMA	20 mm/ 0.3 mm	180 μg/lens	10 mL 0.9% NaCl	24 h			[59]
		MAA	DEAA	EGDMA	14 mm/ 0.08 mm	34.7 μg/lens			180 min	AUC was 8.7 times that of eye drops, MRT was twice as much as eye drops.	[57]
		AA	HEMA	EGDMA	10 mm/ 0.9 mm, 0.2 mm	0.8–3.2 mg/g	2 or 8 mL 0.9% NaCl	0.2 mm 1 day/0.9 mm 14 days		ITC analysis	[60]
		HEMA or DMA	TRIS	EGDMA	7.94 mm/ 0.1 mm	N/A	1 mL PBS (pH 7.4)	6 days		Hyaluronic acid (or PVP) was as a functional additive	[61, 62]
		CmCS, AM	HEMA	EGDMA	10 mm/ 0.1 mm	14–17 μg/ lens	2 mL ALF	4 days		CmCS-g-HEMA-g-PAAM copolymer was synthesized and added in HEMA for the synthesis of lenses	[63]
		MAA	HEMA	EGDMA	N/A	0.35 mg/g	5 mL ALF	12 h		A molecular imprinted structural color contact lens which could self-report the release process by color change	[64]

	Acetazolamide	$ZnMA_2$, HEAA, 1VI or 4VI	HEMA	EGDMA	10 mm/ 0.9 mm	1.22–3.37 mg/ g	5 mL 0.9% NaCl	15 days		Lenses also loaded and released ethoxzolamide (ETOX), which was carbonic anhydrase inhibitor	[40]
		4VI or 1VI, HEAA, Zn^{2+}	NVP, DMA	EGDMA	10 mm/ 0.9 mm	2.47–4.11 mg/ g	10 mL 0.9% NaCl	9 h		Release duration of ETOX exceeds 1 week	[65]
	Dorzolamide	MAA	HEMA	EGDMA	/0.4 mm	N/A	0.9% NaCl, ALF				[41]
Inflammation	Prednisolone acetate	MAA	HEMA	EGDMA	14 mm/ 0.4 mm	58 μg/lens	0.9% NaCl, ALF	2 days			[42]
	Diclofenac sodium	DEAEM	HEMA	PEG200DMA	15 mm/ 105 μm	0.01–0.02 mg/ mg	3 μL/min ALF	6 days		A base curve of 8.6 ± 0.2 mm, a physiological flow by microfluidic chip	[44]
Allergy	Olopatadine	AA, AM, AMPSA, BzMA	HEMA	EGDMA	10 mm/ 0.4 mm	38–82 mg/g	2 or 20 mL ALF	2 days		Histamine and tumor necrosis factor α release were discussed	[49]
	Ketotifen	AA, AM, NVP	HEMA	PEG200DMA	13.5 mm/ 0.4 mm	0.049 mM/g	30 mL ALF (with lysozyme 1 mg/mL)	5 days			[14]
		AA, AM, NVP	HEMA	PEG200DMA	13.5 mm/ 0.4 mm	0.049 mmol/g	3 μL/min ALF	5 days		At a constant zero-order release rate (12.9 μg/day) for about 3.5 days	[51]

(continued)

Table 8.2 (continued)

Category	Drug	Functional monomers	Backbone monomers	Crosslinker	Size (diameter/ tickness)	Drug loading	Release medium	In vitro results	In vivo results	Note	Reference
		AA, AM, NVP	HEMA	PEG200DMA	13.5 mm/ 0.4 mm	N/A	N/A			Transport and structural analysis	[66]
		AA, AM, NVP	HEMA	PEG200DMA	11.8 mm/ 0.1 mm	115 ± 10 μg/ lens	N/A		26 h	AUC was 4365 ± 1070 μg h/ mL, 94 times that of eye drops	[58]

AA acrylic acid, *NVP N*-vinyl pyrrilidone, *HEMA* 2-hydroxyethyl methacrylate, *EGDMA* ethylene glycol dimethacrylate, *HOAc* acetic acid, *TRIS* methacryloxypropyl-*tris*-(trimethylsiloxy) silane, *ALF* artificial lachrymal fluid, *PVP* polyvinylpyrrolidone, *PBS* phosphate buffered saline, *DEAEM* diethylaminoethyl methacrylate, *AM* acrylamide, *DMA* dimethyl acrylamide, *PEG200DMA* poly(ethylene glycol) (200) dimethacrylate, *MAA* methacrylic acid, *MMA* methyl methacrylate, *DEAA N,N*-diethylacrylamide, *CmCS* carboxymethyl chitosan, *ZnMA* zinc methacrylate, *HEAA* hydroxyethyl acrylamide, *1VI* 1-vinylimidazole, *4VI* 4-vinylimidazole, *AMPSA* 2-Acrylamido-2-methyl-1-propanesulfonic acid, *BzMA* benzylmethacrylate, *AUC* the area under concentration-time curve, *MRT* the mean residence time

On March 23, 2021 local time, Johnson & Johnson Vision, a subsidiary of Johnson & Johnson, announced that the Ministry of Health, Labour and Welfare (MHLW) of Japan has officially approved the launch of the anti-allergic (ketotifen) contact lens "Acuvue Theravision." This is the world's first and only drug release contact lens that has been approved by regulatory agencies for marketing. With the launch of the first drug-releasing contact lens, we expect that this innovative technology can be more widely used, bringing more convenient and effective methods for maintaining eye health.

References

1. Hornof M, Toropainen E, Urtti A (2005) Cell culture models of the ocular barriers. Eur J Pharm Biopharm 60:207–225
2. Cunha-Vaz JG (1997) The blood-ocular barriers: past, present, and future. Doc Ophthalmol 93:149–157
3. Duvvuri S, Majumdar S, Mitra AK (2003) Drug delivery to the retina: challenges and opportunities. Expert Opin Biol Ther 3:45–56
4. del Amo EM, Urtti A (2008) Current and future ophthalmic drug delivery systems: a shift to the posterior segment. Drug Discov Today 13:135–143
5. Zhang WS, Prausnitz MR, Edwards A (2004) Model of transient drug diffusion across cornea. J Control Release 99:241–258
6. Peppas NA, Bures P, Leobandung W, Ichikawa H (2000) Hydrogels in pharmaceutical formulations. Eur J Pharm Biopharm 50:27–46
7. Papas EB (2017) Contact lens technology to 2020 and beyond: a review of recent patent literature. Clin Exp Optom 100:529–536
8. Gonzalez-Chomon C, Concheiro A, Alvarez-Lorenzo C (2013) Soft contact lenses for controlled ocular delivery: 50 years in the making. Ther Deliv 4:1141–1161
9. Peng CC, Kim J, Chauhan A (2010) Extended delivery of hydrophilic drugs from silicone-hydrogel contact lenses containing vitamin e diffusion barriers. Biomaterials 31:4032–4047
10. Uchida R, Sato T, Tanigawa H, Uno K (2003) Azulene incorporation and release by hydrogel containing methacrylamide propyltrimenthylammonium chloride, and its application to soft contact lens. J Control Release 92:259–264
11. dos Santos JFR, Alvarez-Lorenzo C, Silva M, Balsa L, Couceiro J, Torres-Labandeira JJ, Concheiro A (2009) Soft contact lenses functionalized with pendant cyclodextrins for controlled drug delivery. Biomaterials 30:1348–1355
12. Yokozaki Y, Sakabe J, Ng B, Shimoyama Y (2015) Effect of temperature, pressure and depressurization rate on release profile of salicylic acid from contact lenses prepared by supercritical carbon dioxide impregnation. Chem Eng Res Des 100:89–94
13. Janagam DR, Wu LF, Lowe TL (2017) Nanoparticles for drug delivery to the anterior segment of the eye. Adv Drug Deliv Rev 122:31–64
14. Venkatesh S, Sizemore SP, Byrne ME (2007) Biomimetic hydrogels for enhanced loading and extended release of ocular therapeutics. Biomaterials 28:717–724
15. Kompella UB, Kadam RS, Lee VHL (2010) Recent advances in ophthalmic drug delivery. Ther Deliv 1:435–456
16. Wilson CG (2004) Topical drug delivery in the eye. Exp Eye Res 78:737–743
17. Holmes B, Brogden RN, Richards DM (1985) Norfloxacin. A review of its antibacterial activity, pharmacokinetic properties and therapeutic use. Drugs 30:482–513
18. Alvarez-Lorenzo C, Yanez F, Barreiro-Iglesias R, Concheiro A (2006) Imprinted soft contact lenses as norfloxacin delivery systems. J Control Release 113:236–244

19. Bedard J, Bryan LE (1989) Interaction of the fluoroquinolone antimicrobial agents ciprofloxacin and enoxacin with liposomes. Antimicrob Agents Chemother 33:1379–1382
20. Marchese AL, Slana VS, Holmes EW, Jay WM (1993) Toxicity and pharmacokinetics of ciprofloxacin. J Ocular Pharmacol 9:69–76
21. Smith A, Pennefather PM, Kaye SB, Hart CA (2001) Fluoroquinolones: place in ocular therapy. Drugs 61:747–761
22. Hui A, Sheardown H, Jones L (2012) Acetic and acrylic acid molecular imprinted model silicone hydrogel materials for ciprofloxacin-HCl delivery. Materials 5:85–107
23. Levin LA, Avery R, Shore JW, Woog JJ, Baker AS (1996) The spectrum of orbital aspergillosis: a clinicopathological review. Surv Ophthalmol 41:142–154
24. Thomas PA (2003) Current perspectives on ophthalmic mycoses. Clin Microbiol Rev 16:730–797
25. Craig JP, Nelson JD, Azar DT, Belmonte C, Bron AJ, Chauhan SK, de Paiva CS, Gomes JAP, Hammitt KM, Jones L, Nichols JJ, Nichols KK, Novack GD, Stapleton FJ, Willcox MDP, Wolffsohn JS, Sullivan DA (2017) TFOS DEWS II report executive summary. Ocul Surf 15:802–812
26. Clayton JA (2018) Dry eye. N Engl J Med 378:2212–2223
27. Bayer IS (2020) Hyaluronic acid and controlled release: a review. Molecules 25:2649
28. Aragona P, Papa V, Micali A, Santocono M, Milazzo G (2002) Long term treatment with sodium hyaluronate-containing artificial tears reduces ocular surface damage in patients with dry eye. Br J Ophthalmol 86:181–184
29. Ali M, Byrne ME (2009) Controlled release of high molecular weight hyaluronic acid from molecularly imprinted hydrogel contact lenses. Pharm Res 26:714–726
30. Buhler N, Haerri HP, Hofmann M, Irrgang C, Muhlebach A, Muller B, Stockinger F (1999) Nelfilcon A, a new material for contact lenses. Chimia 53:269–274
31. Maharana PK, Raghuwanshi S, Chauhan AK, Rai VG, Pattebahadur R (2017) Comparison of the efficacy of carboxymethylcellulose 0.5%, hydroxypropyl-guar containing polyethylene glycol 400/propylene glycol, and hydroxypropyl methyl cellulose 0.3% tear substitutes in improving ocular surface disease index in cases of dry eye. Middle East Afr J Ophthalmol 24:202–206
32. White CJ, McBride MK, Pate KM, Tieppo A, Byrne ME (2011) Extended release of high molecular weight hydroxypropyl methylcellulose from molecularly imprinted, extended wear silicone hydrogel contact lenses. Biomaterials 32:5698–5705
33. Bourne RR, Taylor HR, Flaxman SR, Keeffe J, Leasher J, Naidoo K, Pesudovs K, White RA, Wong TY, Resnikoff S, Jonas JB (2016) Number of people blind or visually impaired by glaucoma worldwide and in world regions 1990-2010: a meta-analysis. PLoS One 11:e0162229
34. Kass MA, Heuer DK, Higginbotham EJ, Johnson CA, Keltner JL, Miller JP, Parrish RK 2nd, Wilson MR, Gordon MO (2002) The ocular hypertension treatment study - a randomized trial determines that topical ocular hypotensive medication delays or prevents the onset of primary open-angle glaucoma. Arch Ophthalmol 120:701–713
35. Garway-Heath DF, Crabb DP, Bunce C, Lascaratos G, Amalfitano F, Anand N, Azuara-Blanco A, Bourne RR, Broadway DC, Cunliffe IA, Diamond JP, Fraser SG, Ho TA, Martin KR, McNaught AI, Negi A, Patel K, Russell RA, Shah A, Spry PG, Suzuki K, White ET, Wormald RP, Xing W, Zeyen TG (2015) Latanoprost for open-angle glaucoma (UKGTS): a randomised, multicentre, placebo-controlled trial. Lancet 385:1295–1304
36. Yan F, Liu YX, Han SL, Zhao QS, Liu NN (2020) Bimatoprost imprinted silicone contact lens to treat glaucoma. AAPS PharmSciTech 21:63
37. Alvarez-Lorenzo C, Hiratani H, Gomez-Amoza JL, Martinez-Pacheco R, Souto C, Concheiro A (2002) Soft contact lenses capable of sustained delivery of timolol. J Pharm Sci 91:2182–2192
38. Hiratani H, Alvarez-Lorenzo C (2002) Timolol uptake and release by imprinted soft contact lenses made of N,N-diethylacrylamide and methacrylic acid. J Control Release 83:223–230

39. Hiratani H, Alvarez-Lorenzo C (2004) The nature of backbone monomers determines the performance of imprinted soft contact lenses as timolol drug delivery systems. Biomaterials 25:1105–1113
40. Ribeiro A, Veiga F, Santos D, Torres-Labandeira JJ, Concheiro A, Alvarez-Lorenzo C (2011) Bioinspired imprinted phema-hydrogels for ocular delivery of carbonic anhydrase inhibitor drugs. Biomacromolecules 12:701–709
41. Malaekeh-Nikouei B, Vahabzadeh SA, Mohajeri SA (2013) Preparation of a molecularly imprinted soft contact lens as a new ocular drug delivery system for dorzolamide. Curr Drug Deliv 10:279–285
42. Malaekeh-Nikouei B, Ghaeni FA, Motamedshariaty VS, Mohajeri SA (2012) Controlled release of prednisolone acetate from molecularly imprinted hydrogel contact lenses. J Appl Polym Sci 126:387–394
43. Moser P, Sallmann A, Wiesenberg I (1990) Synthesis and quantitative structure-activity relationships of diclofenac analogues. J Med Chem 33:2358–2368
44. Tieppo A, Pate KM, Byrne ME (2012) In vitro controlled release of an anti-inflammatory from daily disposable therapeutic contact lenses under physiological ocular tear flow. Eur J Pharm Biopharm 81:170–177
45. Smith DA, Wallwork ML, Zhang J, Kirkham J, Robinson C, Marsh A, Wong M (2000) The effect of electrolyte concentration on the chemical force titration behavior of omega-functionalized SAMs: evidence for the formation of strong ionic hydrogen bonds. J Phys Chem B 104:8862–8870
46. McGill JI (2004) A review of the use of olopatadine in allergic conjunctivitis. Int Ophthalmol 25:171–179
47. Kaliner MA, Oppenheimer J, Farrar JR (2010) Comprehensive review of olopatadine: the molecule and its clinical entities. Allergy Asthma Proc 31:112–119
48. Bilkhu PS, Wolffsohn JS, Naroo SA (2012) A review of non-pharmacological and pharmacological management of seasonal and perennial allergic conjunctivitis. Cont Lens Anterior Eye 35:9–16
49. Gonzalez-Chomon C, Silva M, Concheiro A, Alvarez-Lorenzo C (2016) Biomimetic contact lenses eluting olopatadine for allergic conjunctivitis. Acta Biomater 41:302–311
50. Bielory L (2002) Role of antihistamines in ocular allergy. Am J Med 113:43S–47S
51. Ali M, Horikawa S, Venkatesh S, Saha J, Hong JW, Byrne ME (2007) Zero-order therapeutic release from imprinted hydrogel contact lenses within in vitro physiological ocular tear flow. J Control Release 124:154–162
52. Turner NW, Jeans CW, Brain KR, Allender CJ, Hlady V, Britt DW (2006) From 3D to 2D: a review of the molecular imprinting of proteins. Biotechnol Prog 22:1474–1489
53. Nozik RA, Smolin G, Knowlton G, Austin R (1985) Trimethoprim-polymyxin B ophthalmic solution in treatment of surface ocular bacterial infections. Ann Ophthalmol 17:746–748
54. Zavascki AP, Goldani LZ, Li J, Nation RL (2007) Polymyxin B for the treatment of multidrug-resistant pathogens: a critical review. J Antimicrob Chemother 60:1206–1215
55. Malakooti N, Alexander C, Alvarez-Lorenzo C (2015) Imprinted contact lenses for sustained release of polymyxin B and related antimicrobial peptides. J Pharm Sci 104:3386–3394
56. Hui A, Willcox M, Jones L (2014) In vitro and in vivo evaluation of novel ciprofloxacin-releasing silicone hydrogel contact lenses. Invest Ophthalmol Vis Sci 55:4896–4904
57. Hiratani H, Fujiwara A, Tamiya Y, Mizutani Y, Alvarez-Lorenzo C (2005) Ocular release of timolol from molecularly imprinted soft contact lenses. Biomaterials 26:1293–1298
58. Tieppo A, White CJ, Paine AC, Voyles ML, McBride MK, Byrne ME (2012) Sustained in vivo release from imprinted therapeutic contact lenses. J Control Release 157:391–397
59. Hiratani H, Mizutani Y, Alvarez-Lorenzo C (2005) Controlling drug release from imprinted hydrogels by modifying the characteristics of the imprinted cavities. Macromol Biosci 5:728–733

60. Yanez F, Chauhan A, Concheiro A, Alvarez-Lorenzo C (2011) Timolol-imprinted soft contact lenses: influence of the template: functional monomer ratio and the hydrogel thickness. J Appl Polym Sci 122:1333–1340
61. Guidi G, Korogiannaki M, Sheardown H (2014) Modification of timolol release from silicone hydrogel model contact lens materials using hyaluronic acid. Eye Contact Lens 40:269–276
62. Korogiannaki M, Guidi G, Jones L, Sheardown H (2015) Timolol maleate release from hyaluronic acid-containing model silicone hydrogel contact lens materials. J Biomater Appl 30:361–376
63. Anirudhan TS, Nair AS, Parvathy J (2016) Extended wear therapeutic contact lens fabricated from timolol imprinted carboxymethyl chitosan-g-hydroxy ethyl methacrylate-g-poly acrylamide as a onetime medication for glaucoma. Eur J Pharm Biopharm 109:61–71
64. Deng JZ, Chen S, Chen JL, Ding HL, Deng DW, Xie ZY (2018) Self-reporting colorimetric analysis of drug release by molecular imprinted structural color contact lens. ACS Appl Mater Interfaces 10:34611–34617
65. Ribeiro A, Veiga F, Santos D, Torres-Labandeira JJ, Concheiro A, Alvarez-Lorenzo C (2011) Receptor-based biomimetic NVP/DMA contact lenses for loading/eluting carbonic anhydrase inhibitors. J Membr Sci 383:60–69
66. Venkatesh S, Saha J, Pass S, Byrne ME (2008) Transport and structural analysis for controlled of molecular imprinted hydrogels drug delivery. Eur J Pharm Biopharm 69:852–860

Chapter 9
MIP as Drug Delivery Systems for Special Application

Li Ma and Zhaosheng Liu

9.1 Introduction

The common therapy model seems to moderate success for the majority of patients. Despite the fact that an opportune drug and dosage form is taken in an appropriate manner, quantities of adverse effects are frequently noticed [1–3]. In these considerations, some drug is released immediately and being absorbed in gastrointestinal [4–6]. The regimens of some drug often reach a high peak plasma concentration of the drug, which leads to side effects on the patients, and poor compliance. With the permission of modern therapy, a part of drugs need to be delivered slowly to reduce the side effects, and finally expected therapeutic efficacy [7]. Compared to conventional drug delivery, a drug delivery system (DDS) provides an effective medicinal virtue with lower side effects by controlled release of drugs on significantly localized sites in the body [8, 9]. In current studies, MIPs, which possess selective affinity and sustained-release behavior toward imprinted molecular, have attracted more attention in DDS, especially stimulus-responsive capturing/releasing template under external stimulus such as pH, temperature, as well as magnetic field [10–13]. Therefore, the application of "intelligent" stimulus- responsive MIPs is as carriers of drug in DDS. It is believed that these MIP-based vectors may have the potential to meet the requirements of DDS.

Traditionally, MIPs are prepared by the manner of bulk polymerization as monoliths. However, this method has various defects, including a few of binding sites near to surface, inaccessible recognition sites within the polymer bulk, non-uniform morphology [14, 15]. Besides, large molecules and particularly high-molecular weight macromolecules and biomolecules, such as peptides, proteins, DNA, viruses,

L. Ma
Chengde Food and Drug Inspection and Test Institute, Chengde, Hebei, China

Z. Liu (✉)
College of Pharmacy, Tianjin Medical University, Tianjin, China

Z. Liu et al. (eds.), *Molecularly Imprinted Polymers as Advanced Drug Delivery Systems*, https://doi.org/10.1007/978-981-16-0227-6_9

and bacteria main a formidable challenge to imprint on polymers, on account of their large dimensions, limited solubility and stability, complex structure, slow mass transfer, and structural flexibility in solution [16, 17]. Meanwhile, typical MIPs are synthesized by the polymerization of organic matters. In recent years, MIPs have been combined with inorganic materials, for example, magnetic nanoparticles (MNPs), silica, quantum dots, gold nanoparticles and silver nanoparticles to gain additional properties of the support, e.g., magnetic properties [18]. The development of hybrid materials by combining inorganic and organic materials has the potential to overcome problems mentioned above. Hence, a number of organic-inorganic hybrid nanomaterials have been developed [19, 20].

In particular, the MNPs are widely used as core material with the assist of the technique of core-shell imprinting. Most of the MNPs, with an average diameter of about 20 nm, exhibit superparamagnetic properties and possess the properties of saturation magnetization. They can direct to the targeted location precisely using a week external magnetic field [21, 22]. In addition, MNPs show no magnetization after removal of the magnetic field due to superparamagnetic properties, which is suitable for in vivo applications [23]. These hybrid materials combining the desirable chemical and physical properties of organic and inorganic materials have most obvious advantages over traditional MIPs. On this basis, the application of magnetic molecularly imprinted polymers (MMIPs) has gained considerable attention for the design of hybrid materials attributing to them not only exhibit specific selective binding toward the template molecule but also have outstanding magnetism. The MMIPs, composed of magnetic materials (i.e., iron, nickel, cobalt, or their alloys and oxides) and non-magnetic polymer materials have the characters of magnetic adsorption property, high adsorption capacity to template molecule, and special selective recognition ability [24]. Hence, these MMIPs are suitable for the pre-concentration of analysts as well as for separation and molecular identification of biomolecules, organic and inorganic species in fluidic systems [25]. Moreover, MMIPs can be easily collected/isolated and recycled by the effect of the external magnetic field without additional centrifugation or filtration [26, 27]. MMIPs can also to be directed to specific parts of the body by applying a magnetic field in drug delivery system. The external magnetic field can be induced by a magnet tablet or alternate current. These advantages have showed the possibility of directing MIPs-based drug delivery systems to the targeting area.

Other MIPs applied in DDS have been discussed in previous chapter. Hence, this chapter mainly deals with the design of MIPs as DDS for special application, including MIPs as the carriers in intravenous route, electromagnetic MIPs, and recent use of MIPs as stimuli-responsive and controlled-release drug carrier in most significant application of therapies. Finally, the prospects of MIPs integration with gene therapies will be discussed.

9.2 MIP as Drug Delivery Systems for Intravenous Route

In order to understand the advantage of a controlled release system, the methods of traditional delivery should be discussed. Of these delivery methods, 90–95% are delivered either orally or injected/infused parent rally [28]. A number of drugs are finally resided in the bloodstream and flushed into the circulatory system to perform their therapy. Compared with the oral dosage forms, intravenous injection seem is the most efficient way to subject drug into the blood stream. On the other hand, the bioavailability with the mode of intravenous injections is 100%, which allows for any drug to be delivered with a rapid onset of action. Besides, the investigations of new areas for drug delivery application in molecular imprinting technology need new formats of materials and turn the polymers to nanoscale size. In addition, to avoid drug degradation, MIPs-based DDS has been administrated intravenously by in vitro testing. Hence, the imprinted-based drug delivery devices can be a universal tool for modern therapy.

Recent years, intensive studies of 5-Fluorouracil (5-FU) were carried out as the model drug. As known, 5-FU is an antineoplastic compound that is known to be an intensive chemotherapeutic drug and used broadly against solid tumors and acts by interfering with nucleosides, leading to cytotoxicity and cell death. On account of the rapid rate of metabolism, continuous administration and high serum doses are employed to improve its chemotherapeutic activity in common treatment regimen. However, 5-FU is naturally toxic and enhanced level of concentration causes severe side effects on noncancerous cells [29]. Thus, in order to improve the therapeutic effectiveness and reduce side reactions, the new drug delivery methods are still under exploration.

In early works, the submicroparticles were synthesized using methacrylic acid as functional monomer and ethylene glycol dimethacrylate (EDMA) as crosslinking agent, as a controlled release device for 5-FU in biological fluids [30]. In adsorption and in vitro release experiments, the capacity of the imprinted polymer to bind 5-FU was much more than the corresponding non-imprinted one and showed a controlled/sustained drug release during a period of 30 h. Meanwhile, a 5-FU imprinted cryogel discs with the assistance of Cu^{2+} ion was prepared using metal chelate monomer *N*-methacryloyl-L-histidine [31]. The cumulative release of 5-FU decreased by increasing the cross-linker density in the polymer matrix. However, the above polymers had the following limitations: the submicroscale particles are not appropriate for intravenous injection, impossible for the targeted therapy of cancer cells and polymer matrix is not biodegradable.

To overcome these problems, a new magnetic nanoparticles coated with MIP was fabricated to control the release of 5-FU. Asadi and coworkers [29] developed an effective multi-core-shell structure magnetic molecularly imprinted polymer nanoparticle based on biodegradable materials (tannic acid as a biodegradable monomer, polyphenol as cross-linker agent), as a carrier for targeted, sustained and controlled release of 5-FU. The Fe_3O_4 core provides magnetic properties to nanoparticles which can be used to guide the polymers effectively collected in the disease site under external magnetic field. Such a property of the nano device is very promising in targeted delivery. Additionally, fluoresce in isothiocyanate was incorporated into the nanoparticles to image the location and distribution of the drugs inside the body.

The degradation study of the MIPs nanoparticles was carried out at different conditions including pH 3, 7.4 and 11. The results (Fig. 9.1a) revealed faster degradation at pH 3 or pH 11 than at pH 7.4. 5-FU was immediately released within 2 h in simulated gastric fluid (pH 1.2) and faster degradation. However, the degradation at pH = 7.4 was obviously slower, due to the better sustained and controlled release during a period of 120 h in this condition (Fig. 9.1b). From in vivo studies, the fluorescent images results showed that the MIPs magnetic carriers were effectively collected into the liver under an external magnetic field (Fig. 9.2a–c). Figure 9.2d is a parallel experiment without applying magnet field to evaluate the performance of the magnetic guide. Furthermore, the satisfactory-tumor properties and biocompatibility onto human breast tumor cells were demonstrated by in vitro and in vivo experiment (Fig. 9.2f–j).

The cytotoxicity tests of MIPs sample (without the drug) were measured on NIH/3T3 cell line and the viability percentage of cell has not been demonstrated a dramatic change during 7 days (Fig. 9.2e).

Other 5-FU imprinted nanoparticles were fabricated with the core of Fe_3O_4 coated by a thin layer of polydopamine (PDA) [32]. Hence, the polymerization is possible on the surface of magnetic nanoparticles because the high polarizability, thus, a very thin layer of polymer can be formed on the surface of nanoparticles, without an intermediate layer. Meanwhile, PDA layer can be easily modified and functionalized for different therapies. Another advantage is that the average diameter of particles was approximately 80 nm, which was appropriate for intravenous via the tail vein. In vivo experiments for mouse breast cancer model was made according to the methods that two to three million of murine mammary adenocarcinoma cells (MMAC: derived from M05 cell line) were subcutaneously injected into the flank of female inbred Balb/C mice (6–8 weeks old).

Figure 9.5 shows the tumor growth curves as the relative volumes at different days after treatment. The tumor growth in the free 5-FU treated group and 5-FU imprinted polymers treated group was significantly lower than that in the control and sham groups. In addition, a significant reduction in tumor volume was also observed in the 5-FU imprinted nanoparticles group undermagnetic field compared with all other experimental groups, and the total score of malignancy in the last group was reduced (Fig. 9.3). The above study indicated that targeting 5-FU via magnetic field to tumor site may increase nanodevices local uptake, and the therapeutic efficiency is enhanced. However, the pattern of drug release for 24 h in vitro drug experiment, with an initial burst effect. Within 4 h, 80% 5-FU was released from the nanoparticles. The rapid release of 5-FU could be explained by non-specific adsorption on the surface during the loading of a drug because of the heterogeneous population of binding sites.

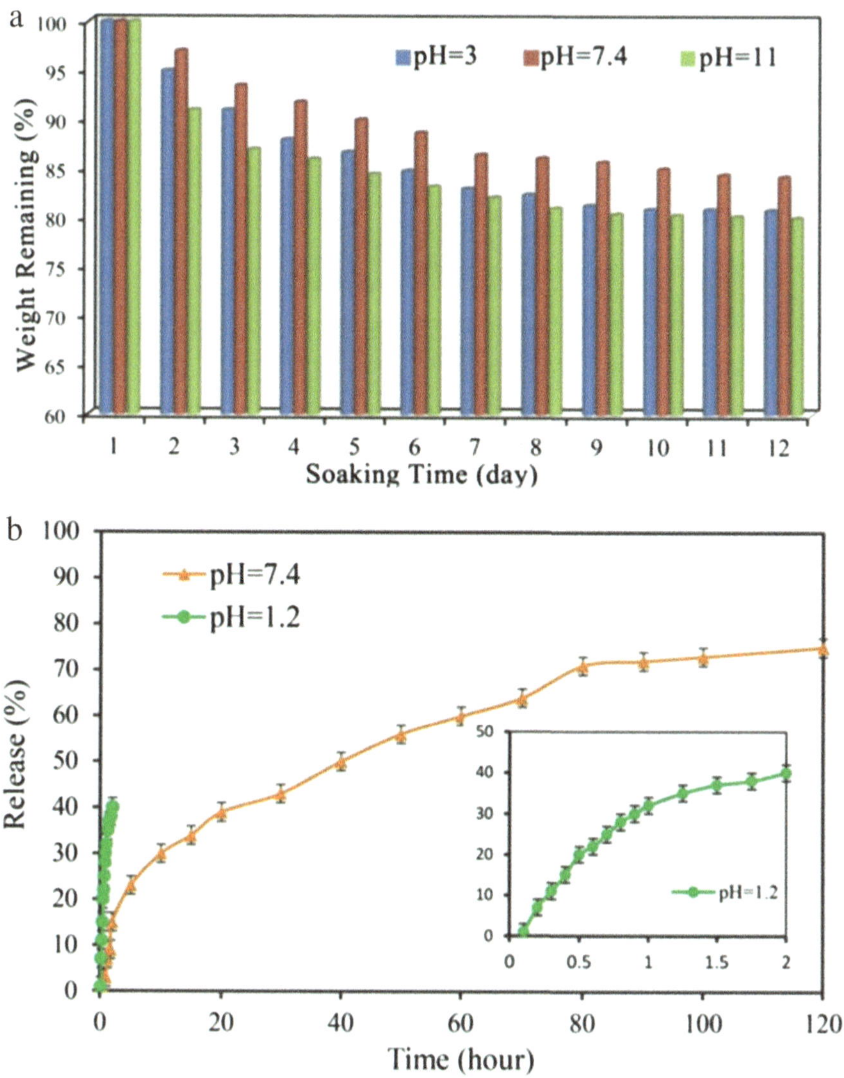

Fig. 9.1 Degradation profile of various pH (3, 7.4 and 11) (**a**) and release profile of 5-FU in simulated gastric fluid (pH = 1.2) and plasma simulating fluid (pH = 7.4), the inset is the magnified profile of drug release at pH = 1.2 for 2 h. Drug release experiments were done triplicate and the results are the average values (**b**). (Reproduced with permission of [29])

In addition to these mentioned above, another MIPs injected into rat brain were prepared based on Olanzapine as template. Olanzapine belongs to the second generation of antipsychotic drugs for the treatments of schizophrenia. The oral administration is the most convenient method for patients. However, around 40% of drug is metabolized before reaching to the systemic circulation and poor

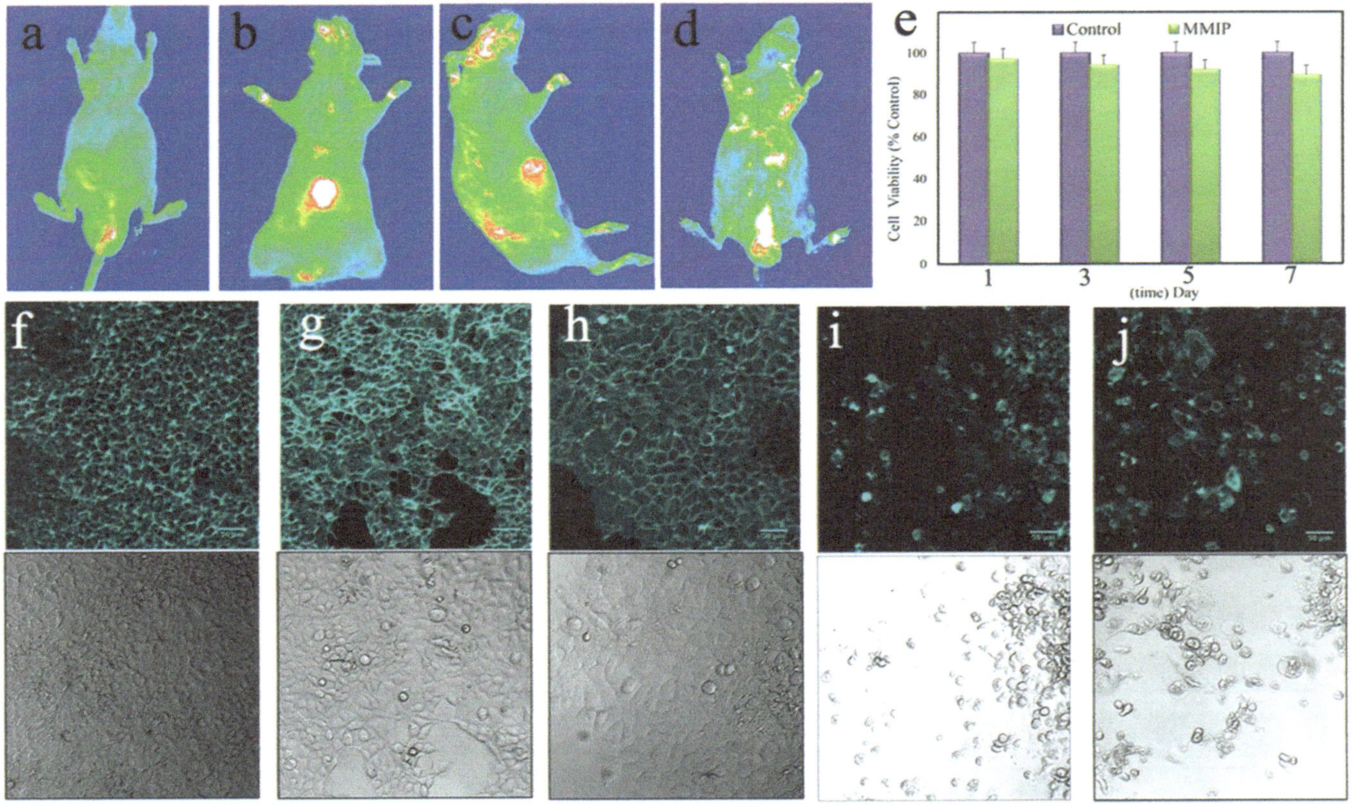

Fig. 9.2 Magnetic-guided drug delivery in rats and related fluorescent image to follow drug delivery to the liver under magnetic field; before injection of carrier (blank) (**a**), 24 h after injection under magnetic field (**b**), 48 h after injection under magnetic field (**c**) and injection without an external magnetic field (**d**). MTT

assay (**e**): effect of MIP nanoparticles on the viability of 3T3 cells. 3T3 NIH cells were incubated with MIP nanoparticles (1, 3, 5 and 7 days), Cell viability assay for assessing the biocompatibility of process and the corresponding images (Bright field images (below) and the fluorescent images (above)); HEK control group (**f**), 24 h after incubation of MMIP nanoparticles (**g**) (48 and 72 h did not show significantly more cell death or changes in morphology), MCF7 control group (**h**), and the anticancer effect of cross linker of tannic acid (**i**) and magnetic molecularly imprinted polymer (**j**) in human breast tumor (MCF-7) cell. (Reproduced with permission of [29])

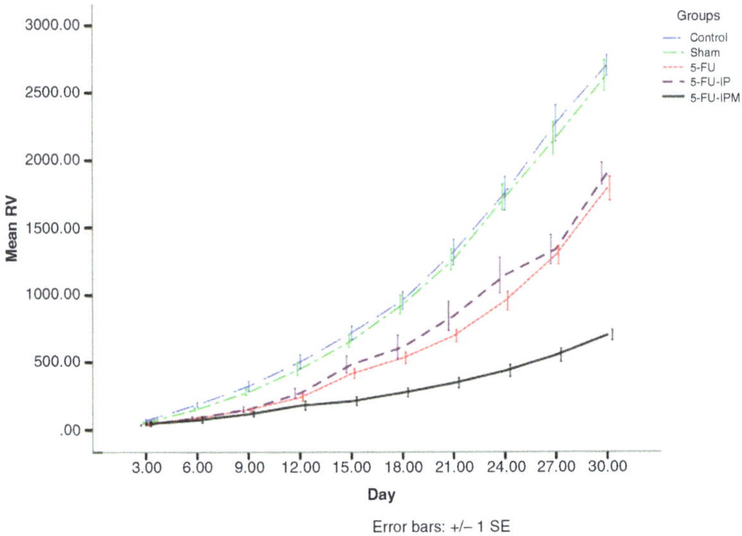

Fig. 9.3 Tumor growth curves of tumor-bearing mice that received different treatments. *5-FU* 5-fluorouracil, *5-FU-IP* 5-fluorouracil imprinted polymer, *5-FU-IPM* 5-fluorouracil imprinted polymer with magnetic field. (Reproduced with permission MPDI publications from of [32])

bioavailability as well as low permeability, limiting its clinical application. To improve therapeutic effect, higher dosage frequencies may causes extra-pyramidal effects, tremors, dry mouth, weight gain, and somnolence. Asadi et al. [33] applied a new biodegradable cross-linker agent for fructose synthesis of an Olanzapine magnetic fluorescent multi-core-shell structured MIPs, as the carriers targeted for brain. The particle in the presence of Olanzapine as template synthesized via coprecipitation polymerization technique was illustrated in Fig. 9.4.

The spherical Fe_3O_4 nanoparticles had average particle size of about 20 nm (Fig. 9.5a–c). Figure 9.5d, e shows the polydispersity index and *z*-average for the MIPs were calculated 0.14 and 58 nm, respectively. When studying the selectivity of the prepared polymer, it was found that the magnetic MIPs possess better affinity to Olanzapine, compared with the solutions of Quetiapine and Clozapine. In addition, the high concentration of fructose was produced from the gradually destruction of the biodegradable carries and the brain cells to use it as fuel, and the loaded drug onto the carrier will be released at the target site. This is a superior property compared with other biodegradable materials such as PEG or lactic acid.

Vinca alkaloids are a group of anti-mitotic and anti-microtubule alkaloid agents originally derived from the periwinkle plant Catharanthusroseus. Vinblastine (VBL), belongs to diindole alkaloids, is one kind of Vinca alkaloids commonly used to treat Hodgkin's lymphoma, non-small cell lung cancer, bladder cancer, brain cancer, testicular cancer, and etc. Nowadays, VBL is administered mainly as injection clinically. Whereas, some clinical problems, for example, the narrow therapeutic

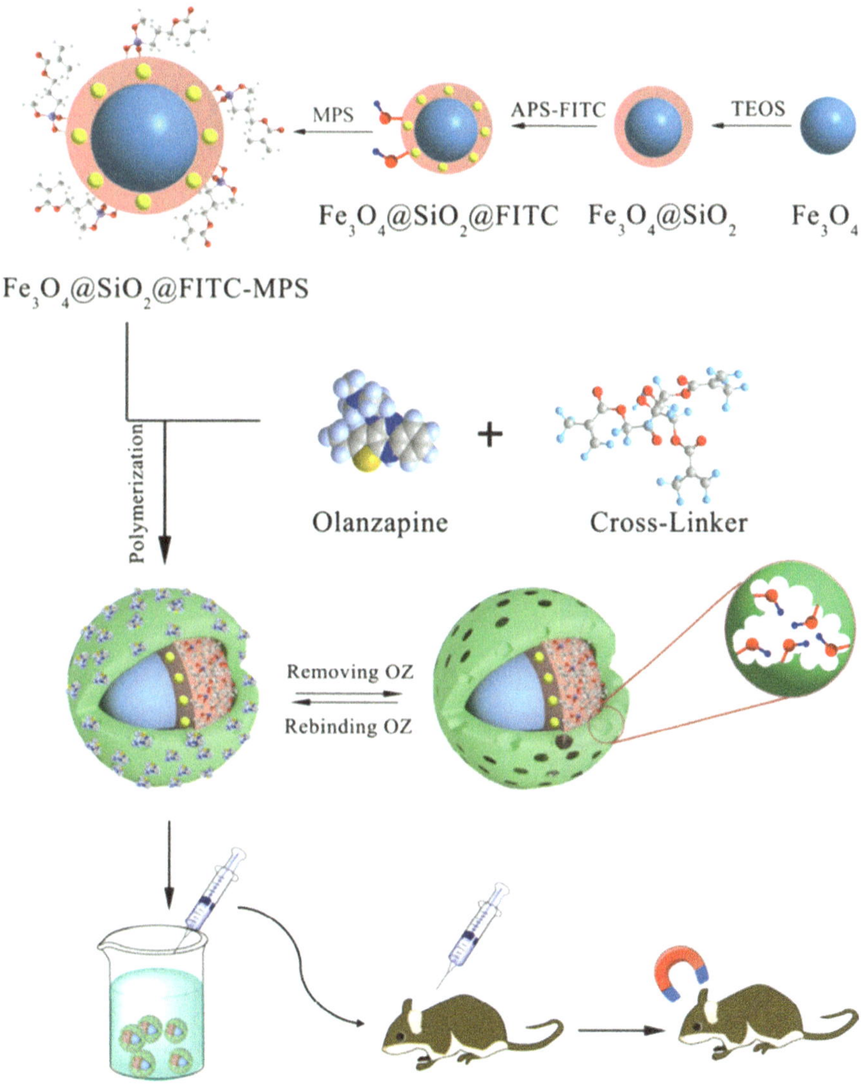

Fig. 9.4 Schematic illustration of the preparation procedure of biodegradable magnetic fluorescent molecularly imprinted polymer for targeting drug delivery of Olanzapine under external magnetic field. (Reproduced with permission of [33])

window and severe dose-limiting, hematological toxicity, including leucopenia, myelosuppression, and anemia, have always restricted its application [33, 34].

To increase antitumor effect and reduce the side effects, nanoparticles maybe an effective way to improve the dilemma, depending on its sustained release and targeting effect. As reported in another paper, the egg sphingomyelin/cholesterol

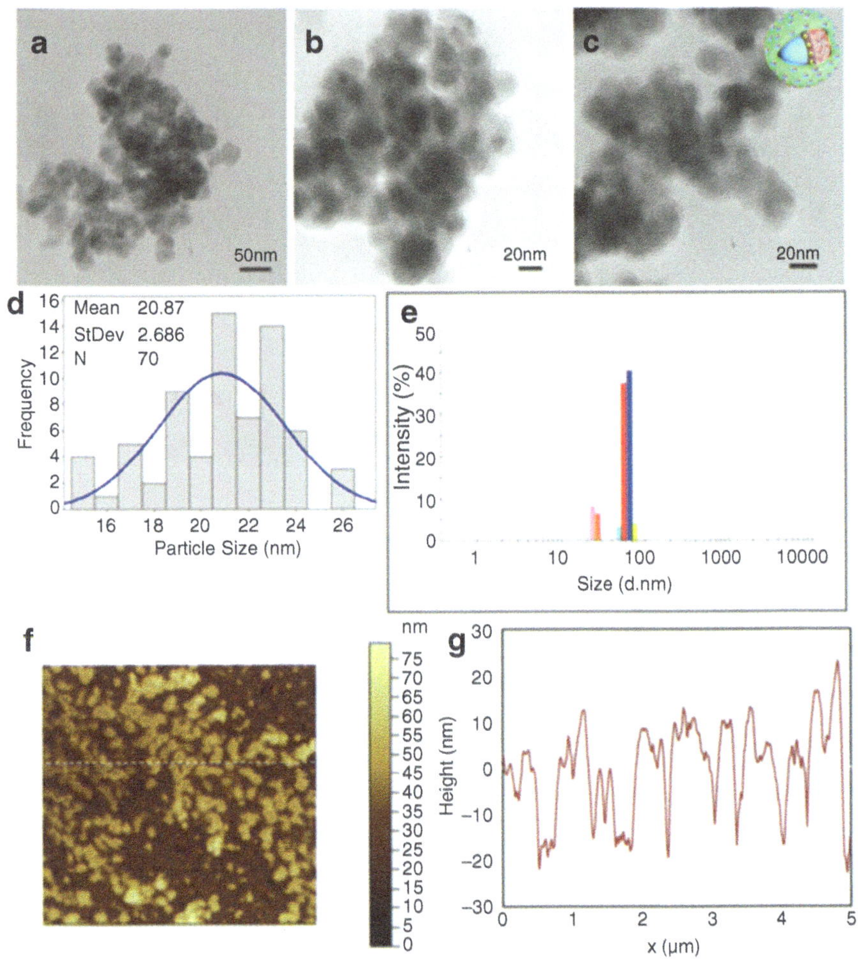

Fig. 9.5 Transmission electron microscope image of nanoparticals: (**a**) Fe_3O_4 nanoparticles (scale bar = 60 nm), (**b**) Fe_3O_4@SiO_2nanoparticles (scale bar = 20 nm), (**c**) Fe_3O_4@SiO_2@MIPs (scale bar = 20 nm), (**d**) the statistical graph of size distribution, standard deviation for Fe_3O_4 nanoparticles, (**e**) dynamic light scattering of MIPs dispersed in acetonitrile, (**f**) atomic force microscope image and (**g**) cross-section profile of the surface of MIPs nanoparticles. (Reproduced with permission of [32])

liposomethe loaded with VBL could achieve sustained release [35]. VBL was loaded in poly (caprolactone) grafted dextran copolymeric nanoparticles which could also sustain release in vitro experiments, as well as enhance cellular uptake of NPs to increase the mortality of cancer cells [36]. Hence, the molecularly imprinted nanoparticles were synthesized by Zhu et al. [37] through precipitation polymerization with VBL as a model drug.

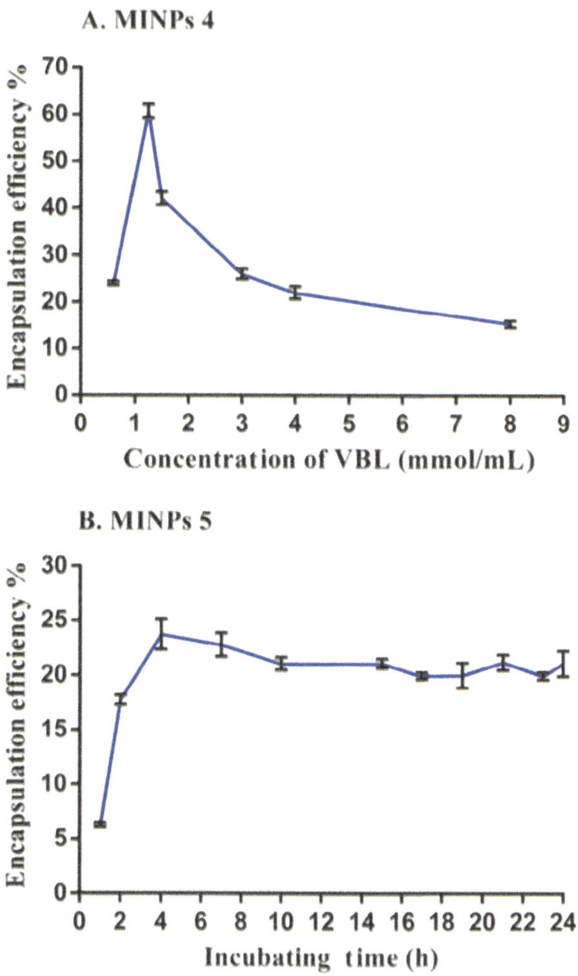

Fig. 9.6 The influence of (**a**) different concentration of VBL and (**b**) adsorption time on EE% of MINPs. (Reproduced with permission of [37])

The different concentration of the VBL solution or different adsorption time influences on nanoparticles loading efficiency. As shown in Fig. 9.6, the EE% of VBL was the highest at the concentration of 1.25 mmol/L. From Fig. 9.6b, it was found that the value reached the top at 4 h and then declined slightly. The loaded VBL adsorbed or precipitating on the surface of the MINPs may result in a quick release. As shown in Fig. 9.7, VBL release from the MINPs-VBL with a burst release in the initial 24 h, which could be explained by the mechanism of solution-diffusion or desorption. Thereafter, the release rate became slow until 216 h, demonstrating a typical behavior of sustained and prolonged drug release.

Bio-distribution study was conducted via injecting the loaded nanoparticles into cauda vein on SD rats. For VBL injection group, the level of VBL was the highest in heart (12.08 ± 0.24μg/mL), followed by kidney, liver, spleen, lung, and the lowest

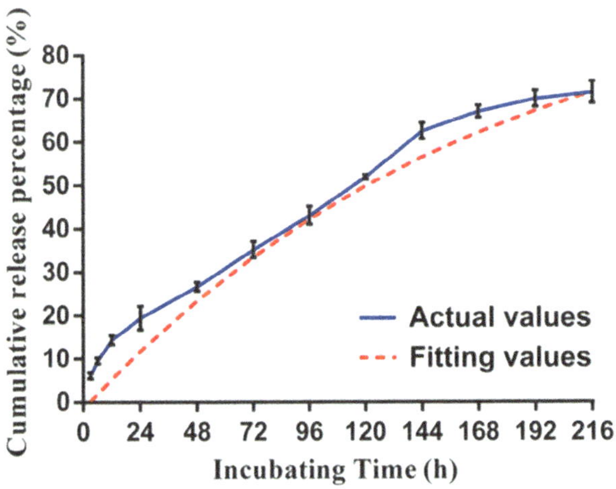

Fig. 9.7 The in vitro release profile of MINPs-VBL. (Reproduced with permission of [37])

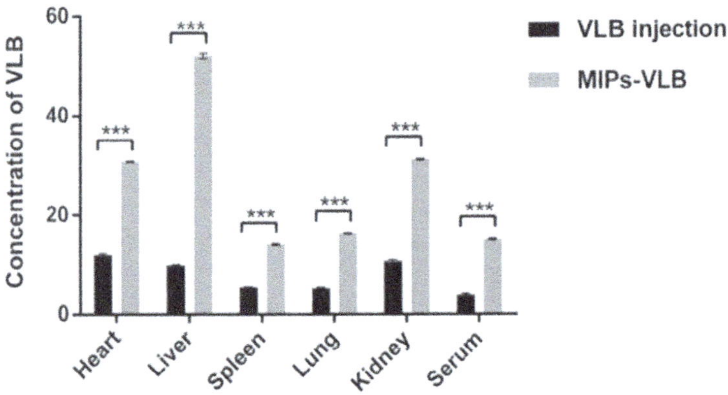

Fig. 9.8 Concentration of VBL in serum and different tissues of rats after tail i.v. of MINPs-VBL and VBL injection ($n = 6$). (Reproduced with permission of [37])

in serum (Fig. 9.8). Compared with the VBL injection group, the drug level of the MINPs-VBL was higher than VBL injection group. It suggested that the sustained-release property of the MINPs contributed to the higher drug levels of the MINPs-VBL in tissues and serum.

9.3 Electromagnetic MIPs for Drug Delivery Systems

Compared with other magnetic materials, for example, nickel, cobalt, or their alloys and oxides, magnetite Fe_3O_4 is the most widely used magnetic material in biomedicine and biotechnology, due to ease of preparation, magnetism biocompatibility and the low toxicity [38]. It was approval by the Food and Drug Administration as the only type of MNPs for clinical usage [39]. For a decade, the common method for MNPs preparation included co-precipitation, solvothermal/hydrothermal and thermal decomposition. In general, these magnetic nanoparticles with an average diameter of about 20 nm made with common method of generate heat when exposed to an alternate current magnetic field due to the relaxation of its magnetic moment. The ability of heat has attracted considerable attention for magnetic hyperthermia applications [40, 41]. Moreover, the amount of heat is related to the strength of magnetic field and the size of magnetic nanoparticles [42, 43]. Hence, magnetite nanoparticles have potential to be applied in cancer therapy using magnetic hyperthermia as a magnetic thermal seed [44].

The power of AC magnetic field can be controlled by the strength of electric current. However, it is considered difficultly to kill cancers cells through heat therapy only by hyperthermia effect. As well, the heat therapy needs a mass of nanoparticles to provide macroscopic heat to achieve therapeutic, it is quite unattainable. Therefore, anticancer drug treatment is indispensable in concurrent therapy. Consequently, a novel DDS carrier which can achieve directivity toward the targeting area is developed. Particularly, the amount of magnetically triggered nanomaterials has been reported for new stimulus drug releasing [45, 46].

Recently, most of the MMIPs studies applied in DDS in present of magnetic field were concentrated on cells or in vitro tests. A new MIPs coated on MNPs with PDA modified was employed for the controlled release of Doxorubicin (DOX) [47]. Higher efficacy of the DOX-imprinted PDA (DOX-IP) with magnetic field in suppressing tumor growth than free DOX and DOX-IP without magnetic field was obtained. Other similar results were exhibited by Hamid's group [32]. They aimed to investigate the controlled delivery of 5-FU in the treatment of a mouse breast cancer model undermagnetic field. 5-FU-IP in the presence of magnetic field behaved superior treatment than DOX-IP without magnetic field.

A novel hybrid molecularly imprinted poly (acrylamide) shell around γ-Fe_2O_3 core was made [48]. The core γ-Fe_2O_3 NPs were firstly synthesized by a method of coprecipitation, and a size sorting process through salt destabilization was followed to get the biggest NPs which was beneficial for magnetic hyperthermia [49]. The particles with an average particle diameter (d_0) of 11 nm and a polydispersity (σ) of 0.31 according to TEM analysis. To stabilize and functionalize the Fe_2O_3 NPs, the NPs further cross-linked with EDMA was presented by Griffete [48]. The doxorubicin (DOX) was imprinted in situ during the process of polymerization reaction and held by the poly (acrylamide) mesh via H-bonding (Fig. 9.9).

The authors investigated the potential application of the Fe_2O_3@DOX-MIP for drug delivery. The in vitro release studies were monitored in various conditions: in a

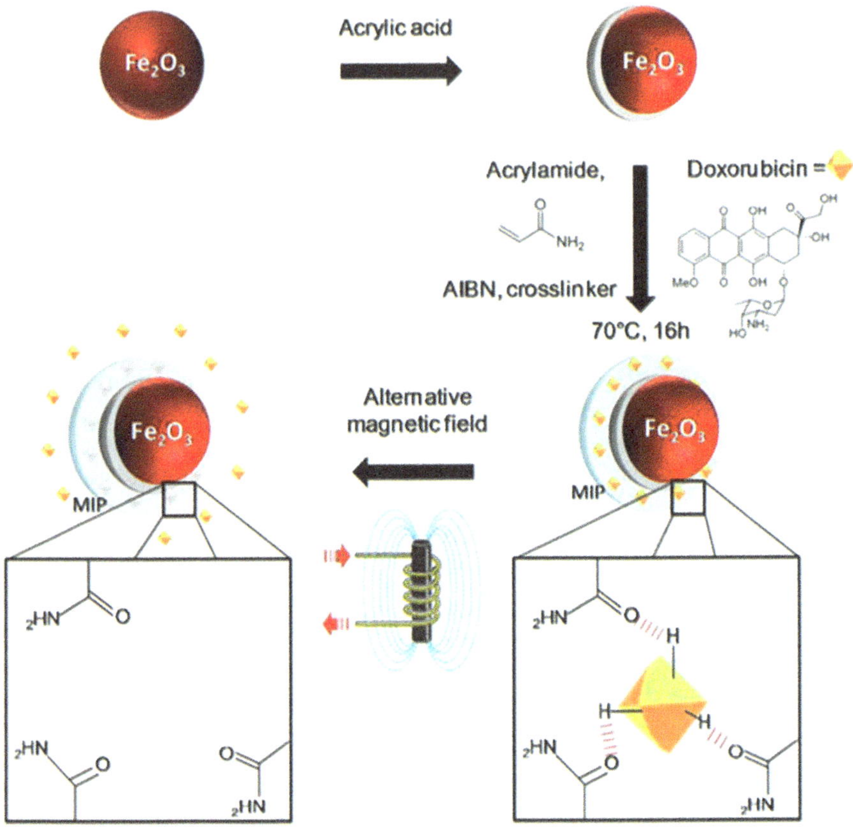

Fig. 9.9 Preparation and DOX loading in MIP γ-Fe_2O_3NPs. (Reproduced with permission of [48])

water bath at macroscopic T maintained temperature (37 °C) close to human body and under the "athermal" an AMF (alternate current magnetic field) (5 AMF pulses of 2 min with a 30 s interval, 335 kHz, 9 mT) induced drug released at 37 °C. The result showed that the 60% of DOX loaded on the Fe_2O_3@DOX-MIP was released with applied AMF, which was four times higher drug quantity than a sample left for 8 h at 37 °C without magnetic field. For each point of the release curve, the amount of DOX was significantly higher when the nanoparticles were submitted to an AC magnetic field. The reason may be that the hydrogen bonds between the MIPs and DOX are broken and the molecule is released due to AC magnetic field, at the same time, the medium without any significant heating.

The particles were submitted to the same AC magnetic field excitation for the Fe_2O_3@DOX-MIP NPs. A high DOX concentration release was achieved after a sample left for 8 h at 37 °C. With the effect of AC magnetic field, the same material showed the complete release of DOX after 8 h. In both Fe_2O_3@DOX-MIP NPs and Fe_2O_3@NIP NPs, the slight effect observed upon AC magnetic field excitation is probably due to a more rapid diffusion of DOX molecules with the temperature elevation in the polymer matrix. Besides, the Fe_2O_3@DOX-MIP nanoparticles were heated at different temperatures to evaluate the temperature effect on the drug release profile.

The results show that the DOX concentration released at 37 °C is three times lower than the amount of DOX released after 4 h heating in water with AMF pulse at each time point under the same temperature. PC3 cancer cells did not die after internalized these Fe_2O_3@DOX-MIP nanoparticles, due to inactive DOX. By contrast, after AMF application R1–2 (at 700 kHz, 25 mT), in isothermal conditions for $t = 30$ min, 1.5 h, and 2.5 h for AMF treatment, the viability of cancer cell was reduced to 60% after 1.5 h treatment.

In another paper, a hybridizing thermal-responsive MIPs with Fe_3O_4 nanoparticle as the carrier for an anticancer drug, methotrexate (MTX) has been proposed [13]. This MIPs thermosensitive DDS (MIPs-MTS) not only generates heat but also releases drug simultaneously. The Fe_3O_4-based magnetic nanoparticles were prepared by the coprecipitation method followed by modification of vinyl groups (VTMS-MTS). The average diameter of the particles was estimated to be 11 nm. Then thermal-responsive MIPS for MTX was coated onto the prepared VTMS-MTS. Folic acid was employed as a pseudo template molecule, methacrylic acid (MAA) as a functional monomer and then crosslinked by divinylbenzene (DVB). Several of characterization methods, including FT-IR spectra, TEM, DLS, have been conducted to confirm the successful modification of MIPs layer onto VTMS-MTS. Meanwhile, original MTS and other hybrid particles showed S shaped curves and disappearance of hysteresis, suggesting superparamagnetic behavior. The selective adsorption of the MIP-MTS and ncMIPs were confirmed against corresponding NIP. Interestingly, both the amount and selectivity for MTX were dramatically increased in MIP-MTS since the recognition sites were rigidly constructed on the surface of the nanoparticles. Moreover, the MIPs-MTS exhibited a slight enhancement in selective recognition of MTX, compared with the compounds having structures with the similar moiety. The increasing release amount and ratio of the desorbed MTX was obtained in both MIPs and NIPs at higher temperatures. As expected, the desorption ratio of the NIPs was higher than that of the MIPs at 60 °C due to MTX adsorbed on the NIP weakly through hydrogen bonding, and 60 °C was enough to cleave most of the interactions. It was reported in another paper that the release rate of 2,4,5-trichlorophenol from a thermo-responsive MMIPs was increased with the increasing of temperature from 25 to 60 °C [50]. Additionally, these magnetic nanoparticles generate heat under AC magnetic field which is related to the concentration of MTS. The above results are confirmed that the hybridizing MIP-MTS exhibits excellent thermo-responsive performance. Considering the results related to selective adsorption/desorption experiment and heating function in the MIP-MTS, finally the drug released through applying AC magnetic field to stimulate MTX release. The difference in the amount of MTX released was clearly confirmed in the presence or absence of AC magnetic field (Fig. 9.10). These Fe_3O_4 particles were heated by magnetic field, and then MTX loaded on the MIP-MTS was effectively cleaved from imprinted sites. The amount of MTX increased remarkably with the increasing concentration of the MIP-MTS (Fig. 9.10a). However, the heating plateaued was

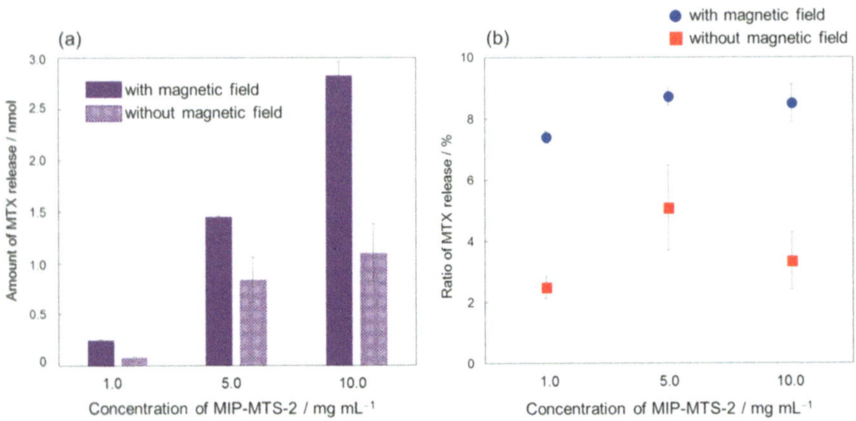

Fig. 9.10 Releasing MTX from MIP-MTS-2 by AC magnetic field. (**a**) Amount and (**b**) ratio of released MTX from MIP-MTS-2 with/without AC magnetic field. The ratio was estimated from the adsorbed MTX on MIP-MTS-2 in advance. (Reproduced with permission of [13])

gained at higher concentration of the MIP-MTS due to the finite power of AC magnetic field. Consequently, the ratio of MTX released was almost same from 5 mg/mL of the MIP-MTS to 10 mg/mL (Fig. 9.10b). The above results demonstrated that the possibility of using hybrid thermal-responsive MIPs and magnetic thermal seed for efficient drug release as a new DDS.

9.4 MIP as Drug Delivery Systems of Gene Therapies

Gene therapy which introduces the exogenous gene into target cells has long fascinated scientists, clinicians, and the general public because of the superior potential to treat a disease by genetic roots. The future of medicine moves toward the development of a lifelong steady incremental care and biotechnologies that produce gene-altering therapeutics [51–53]. After 2015, gene therapy entered a new development phase, and gene therapy drugs were officially authorized by regulatory agencies every year [54]. Until August 2019, 22 gene medicines had been approved by the drug regulatory agencies from various countries [55]. FDA has recently approved Zolgensma (Onasemnogene Abeparvovec-xioi), an adeno-associated viral vector-based gene therapy drug for pediatric spinal muscular atrophy (SMA) patients [56].

For instance, the most frequently reported adverse reactions were elevated by alanine aminotransferase (ALT) and/or aspartate aminotransferase (AST) levels above the upper limit of normal [57]. Mendell et al. [58] also noted other complications, including respiratory illness, as well as clinically asymptomatic transient elevations in ALT and AST levels. For instance, poor pharmacokinetic properties are probably thought to limit the efficacy of these formulations, which include low stability in the circulation, poor tissue-targeting ability and degradation in lysosomes [59].

It is worth noting that the adeno-associated viral vectors have been used to deliver gene in clinic trials [60]. Moreover, the basic defect problems, for example, the questions of application in immunology, insertional mutagenesis, and the toxic effect of viruses and the capacity of gene and so on, bring difficulties during the period of treatment [61]. Nonviral carriers have several advantages over the viral counter parts, due to a higher amount of cargo, lower manufacturing cost, and low to no immunogenicity [62]. Alongside with poly-(lactide-*co*-glycolide), dendritic macromolecule [63, 64] and liposome [65] have been demonstrated to be able to take the place of viral carriers as gene delivery. Besides, inorganic nanoparticles [66], PEGylation [67] and peptides were also used as gene vector in current research [68].

With the development of robust MIPs as carriers of macromolecule (protein [64, 69, 70], nucleic acids [71, 72], and *Staphylococcus aureus* [73]), for the detection of template, solid-phase extraction, as well as biomimetic sensors, may act as a systemic circulation delivery agent. Particularly, the multi-walled carbon nanotubes MIPs with guanine sites of DNA as recognition element to determine G-rich DNA in the human urine and human serum (1%) samples have been demonstrated by Min et al. [74]. The schematic illustration of the assembly procedures for the MIPs composites and the electrochemical detection are indicated.

In another group, Minoura et al. demonstrated the feasibility of MIPs as a selective DNA recognition and trapping polymer [75]. Furthermore, a electropolymerized hexameric 2,2′-bithien-5-yl oligonucleotide probe, immobilized in MIPs cavities, is proven robust and sufficiently specific to detect one nucleo base mismatch at room temperature within 2 min [76].

In particular, pathological tumor sites exhibit local biochemical abnormalities, which can be used to trigger and activate drug release [77]. Due to the rapid development of stimuli-responsive MIPs [13, 78], magnetic-responsive style [48, 79], dual- and multi-responsive MIPs [80] and so on, the robust pH-triggered release mechanism in MIPs is particularly suitable for the target delivery of payload to the acidic tumor microenvironment. The pH value of cancer cells intracellular compartment is lower than the extracellular environment, due to the anisotropic extra cellular acidosis [81], which may potentially assist polymeric materials like MIPs to release its payload (e.g., tumor-associated antigen) and be dependent on pH extra cellularly to help attract T cells to the desired tumor sites.

Moreover, MIPs are generally described as synthetic analogues to the natural, biological antibody-antigen systems, and their tailor-made molecular cavities have been described as "plastic antibodies." As such, these MIPs selectively bind the molecular template during production by the means of a "lock and key" mechanism. Chianella et al. [82] modified the common enzyme-linked immunosorbent assay by direct replacement of antibodies with MIPs. Liu et al. [83] recently utilized MIPs based artificial antibodies for intracellular proteins. At the same time, the antigen/epitope imprinting technologies were specially applied to synthesize MIPs based on HER2 epitope, imprinting on silica nanoparticles, for targeted drug delivery in ovarian cancer mouse model [84], and an inducible epitope imprinting strategy was put to assist MIPs remold the original peptide into the expected conformation and specifically bind to the corresponding protein [85]. Further, some scFvs, modified by the way of protein engineering and chemical substitution, can be conjugated to nanoparticles and polymers via the cysteine residues of its peptide linker [86] and by virtue of reducing the modified cysteinethiol [87]. As mentioned above, MIPs are probably as the carriers of multiple antibodies or antibody-like functionalities to give assistance to drug delivery systems with sustained and environment-responsive drug release mechanisms.

9.5 Conclusions

The above investigations for imprinted nanoparticles show progress in drug delivery devices with huge advantages of targeted delivery and nanoscale. The employment of these natural materials can be helpful in fitting the regulatory requirements for a new drug delivery. However, the application of drug vehicles in DDS just reaches nascent stages. For example, the biocompatibility and clinical trials have not confirmed. Many interesting studies have been focused on in vitro experiments, such as electromagnetic MIPs. Finally, with a tremendous leap of gene therapies, particularly in the past two decades, MIPs seem to be a promising role to deliver gene therapies, the room of improvements of MIPs to translate gene is anticipated.

References

1. Cohen JS (1999) Ways to minimize adverse drug reactions. Individualized doses and common sense are key. Postgrad Med 106:163–172
2. Jara AJ, Zamora MA, Skarmeta AF (2014) Drug identification and interaction checker based on IoT to minimize adverse drug reactions and improve drug compliance. Pers Ubiquitous Comput 18:5–17
3. Clark WG, Brater DC, Johnson AR (1988) Goth's medical pharmacology. The CV Mosby Company, St. Louis
4. Ravic M, Warrington S, Boyce M, Dunn K, Johnston A (2004) Repeated dosing with donepezil does not affect the safety, tolerability or pharmacokinetics of single-dose thioridazine in young volunteers. Br J Clin Pharmacol 58:34–40
5. Mahmood I (2010) Theoretical versus empirical allometry: facts behind theories and application to pharmacokinetics. J Pharm 99:2927–2933
6. Esfandyari-Manesh M, Javanbakht M, Dinarvand R, Atyabi F (2012) Molecularly imprinted nanoparticles prepared by miniemulsion polymerization as selective receptors and new carriers for the sustained release of carbamazepine. J Mater Sci Mater Med 23:963–972
7. Florence AT, Lee VHL (2011) Personalised medicines: more tailored drugs, more tailored delivery. Int J Pharm 415:29–33
8. Dissanayake S, Denny WA, Gamage S, Sarojini V (2017) Recent developments in anticancer drug delivery using cell penetrating and tumor targeting peptides. J Control Release 250:62–76
9. Karimi M, Ghasemi A, Sahandi Zangabad P, Rahighi R, Moosavi Basri SM, Mirshekari H, Amiri M, Shafaei Pishabad Z, Aslani A, Bozorgomid M, Ghosh D, Beyzavi A, Vaseghi A, Aref

AR, Haghani L, Bahrami S, Hamblin MR (2016) Smart micro/nanoparticles in stimulus-responsive drug/gene delivery systems. Chem Soc Rev 45:1457–1501
10. Culver HR, Clegg JR, Peppas NA (2015) Analyte-responsive hydrogels: intelligent materials for biosensing and drug delivery. Acc Chem Res 50:170–178
11. Koetting MC, Peters JT, Steichen SD, Peppas NA (2015) Stimulus-responsive hydrogels: theory, modern advances, and applications. Mater Sci Eng R 93:1–49
12. Li W, Dong K, Ren JS, Qu XG (2016) A beta-lactamase-imprinted responsive hydrogel for the treatment of antibiotic-resistant bacteria. Angew Chem 55:8049–8053
13. Kubo T, Tachibana K, Naito T, Mukai S, Akiyoshi K, Balachandran J, Otsuka K (2019) Magnetic field stimuli-sensitive drug release using a magnetic thermal seed coated with thermal-responsive molecularly imprinted polymer. ACS Biomater Sci Eng 5:759–767
14. Ding XC, Heiden PA (2014) Recent developments in molecularly imprinted nanoparticles by surface imprinting techniques. Macromol Mater Eng 299:268–282
15. Rutkowska M, Potka-Wasylka J, Morrison C, Wieczorek PP, Namieśnik J, Marć M (2018) Application of molecularly imprinted polymers in analytical chiral separations and analysis. Trends Anal Chem 102:91–102
16. Dinc M, Basan H, Hummel T, Müller M, Sobek H, Rapp I, Diemant T, Behm RJ, Lindén M, Mizaikoff B (2018) Selective binding of inhibitor-assisted surface-imprinted core/shell microbeads in protein mixtures. ChemistrySelect 3:4277–4282
17. Zahedi P, Ziaee M, Abdouss M, Farazin A, Mizaikoff B (2016) Biomacromolecule template-based molecularly imprinted polymers with an emphasis on their synthesis strategies: a review. Polym Adv Technol 27:1124–1142
18. Niu M, Pham-Huy C, He H (2016) Core-shell nanoparticles coated with molecularly imprinted polymers: a review. Microchim Acta 183:2677–2695
19. Xu CG, Shen XT, Ye L (2012) Molecularly imprinted magnetic materials prepared from modular and clickable nanoparticles. J Mater Chem 22:7427–7433
20. Zhang ZJ, Zhang XH, Liu BW, Liu JW (2017) Molecular imprinting on inorganic nanozymes for hundred-fold enzyme specificity. J Am Chem Soc 139:5412–5419
21. Presa PDL, Luengo Y, Multigner M, Costo R, Morales MP, Rivero G, Hernando A (2015) Study of heating efficiency as a function of concentration, size, and applied field in γ-Fe2O3 nanoparticle. J Phys Chem C 116:25602–25610
22. Kolosnjaj-Tabi J, Di Corato R, Lartigue L, Marangon I, Guardia P, Silva AKA, Luciani N, Clement O, Flaud P, Singh JV, Decuzzi P, Pellegrino T, Wilhelm C, Gazeau F (2014) Heat-generating iron oxide nanocubes: subtle "destructurators" of the tumoral microenvironment. ACS Nano 8:4268–4283
23. Dinc M, Esen C, Mizaikoff B (2019) Recent advances on core-shell magnetic molecularly imprinted polymers for biomacromolecules. Trends Anal Chem 114:202–217
24. Ansari S, Karimi M (2017) Recent configurations and progressive uses of magnetic molecularly imprinted polymers for drug analysis. Talanta 167:470–485
25. Aguilar-Arteaga K, Rodriguez J, Barrado E (2010) Magnetic solids in analytical chemistry: a review. Anal Chim Acta 674:157–165
26. Li JY, Zu BY, Zhang Y, Guo XZ, Zhang HQ (2010) One-pot synthesis of surface-functionalized molecularly imprinted polymer microspheres by iniferter-induced "living" radical precipitation polymerization. J Polym Sci Polym Chem 48:3217–3228
27. Wu XQ, Wang XY, Lu WH, Wang XR, Li JH, You HY, Xiong H, Chen L (2016) Water-compatible temperature and magnetic dual-responsive molecularly imprinted polymers for recognition and extraction of bisphenol A. J Chromatogr A 1435:30–38
28. Kuhn LT (2018) Introduction to biomedical engineering, Biomaterials, vol 8. Academic Press, New York, pp 219–271
29. Asadi E, Abdouss M, Leblanc R, Ezzati N, Wilson J, Azodi Deilami S (2016) In vitro/in vivo study of novel anti-cancer, biodegradable cross-linked tannic acid for fabrication of 5-fluorouracil-targeting drug delivery nano-device based on a molecular imprinted polymer. RSC Adv 6:37308–37318

30. Puoci F, Iemma F, Cirillo G, Picci N, Matricardi P, Alhaiqu F (2007) Molecularly imprinted polymers for 5-fluorouracil release in biological fluids. Molecules 12:805–814
31. Cetin K, Denizli A (2015) 5-Fluorouracil delivery from metal-ion mediated molecularly imprinted cryogel discs. Colloids Surf B Biointerfaces 126:401–406
32. Hashemi-Moghaddama H, Kazemi-Bagsangania S, Mamilib M, Zavarehc S (2016) Evaluation of magnetic nanoparticles coated by 5-fluorouracil imprinted polymer for controlled drug delivery in mouse breast cancer model. Int J Pharm 497:228–238
33. Marinina J, Shenderova A, Mallery SR, Schwendeman SP (2000) Stabilization of vinca alkaloids encapsulated in poly(lactide-co-glycolide) microspheres. Pharm Res 17:677–683
34. Zu YG, Zhang Y, Zhao XH, Zhang Q, Liu Y, Jiang R (2009) Optimization of the preparation process of vinblastine sulfate (VBLS)-loaded folate-conjugated bovine serum albumin (BSA) nanoparticles for tumor-targeted drug delivery using response surface methodology (RSM). Int J Nanomedicine 4:321–333
35. Zhigaltsev IV, Maurer N, Akhong QF, Leone R, Leng E, Wang JF (2005) Liposome-encapsulated vincristine, vinblastine and vinorelbine: a comparative study of drug loading and retention. J Control Release 104:103–111
36. Prabu P, Chaudhari A, Dharmaraj N, Khil MS, Park SY, Kim HY (2009) Preparation, characterization, in-vitro drug release and cellular uptake of poly(caprolactone) grafted dextran copolymeric nanoparticles loaded with anticancer drug. J Biomed Mater Res A 90:1128–1136
37. Zhu YY, Yang L, Huang DD, Zhu QH (2017) Molecularly imprinted nanoparticles and their releasing properties, bio-distribution as drug carriers. Asian J Pharm Sci 2:172–178
38. Płotka-Wasylka J, Szczepańska N, de la Guardia M, Namiessnik J (2016) Modern trends in solid phase extraction: new sorbent media. Trends Anal Chem 77:23–43
39. Figuerola A, Corato RD, Manna L, Teresa P (2010) From iron oxide nanoparticles towards advanced iron-based inorganic materials designed for biomedical applications. Pharmacol Res 62:126–143
40. Fortin JP, Wilhelm C, Servais J, Ménager C, Bacr JC, Gazeau F (2007) Size-sorted anionic iron oxide nanomagnets as colloidal mediators for magnetic hyperthermia. J Am Chem Soc 129:2628–2635
41. Yu MK, Jeong YY, Park J, Park S, Kim JW, Min JJ, Kim K, Jon S (2010) Drug-loaded superparamagnetic iron oxide nanoparticles for combined cancer imaging and therapy in vivo. Angew Chem Int Ed 47:5362–5365
42. Sadat ME, Patel R, Sookoor J, Bud'ko SL, Ewing RC, Zhang J, Xu H, Wang Y, Pauletti GM, Mast DB, Shi D (2014) Effect of spatial confinement on magnetic hyperthermia via dipolar interactions in Fe3O4 nanoparticles for biomedical applications. Mater Sci Eng C 42:52–63
43. Suto M, Hirota Y, Mamiya H, Fujita A, Kasuya R, Tohji K, Jeyadevan B (2009) Heat dissipation mechanism of magnetite nanoparticles in magnetic fluid hyperthermia. J Magn Magn Mater 321:1493–1496
44. Di Corato R, Béalle G, Kolosnjaj-Tabi J, Espinosa A, Clément O, Silva AK, Ménager C, Wilhelm C (2015) Combining magnetic hyperthermia and photodynamic therapy for tumor ablation with photoresponsive magnetic liposomes. ACS Nano 9:2904–2916
45. Hoare T, Timko BP, Santamaria J, Goya GF, Irusta S, Lau S, Stefanescu CF, Lin D, Langer R, Kohane DS (2011) Magnetically triggered nanocomposite membranes: a versatile platform for triggered drug release. Nano Lett 11:1395–1400
46. Omar H, Croissant JG, Alamoudi K, Alsaiari S, Alradwan I, Majrashi MA, Anjum DH, Martins P, Laamarti R, Eppinger J, Moosa B, Almalik A, Khashab NM (2017) Biodegradable magnetic silica@ iron oxide nanovectors with ultra-large mesopores for high protein loading, magnetothermal release, and delivery. J Control Release 259:187–194
47. Zavareh S, Mahdi M, Erfanian S, Hashemi-Moghaddam H (2016) Synthesis of polydopamine as a new and biocompatible coating of magnetic nanoparticles for delivery of doxorubicin in mouse breast adenocarcinoma. Cancer Chemother Pharmacol 78:1073–1085

48. Griffete N, Fresnais J, Espinosa A, Wilhelm C, Bée A, Ménager C (2015) Design of magnetic molecularly imprinted polymer nanoparticles for controlled release of doxorubicin under alternative magnetic field in athermal conditions. Nanoscale 7:18891–18896
49. Lefebure S, Dubois E, Cabuil V, Neveu S, Massart R (1998) Monodisperse magnetic nanoparticles: preparation and dispersion in water and oils. J Mater Res 13:2975–29781
50. Wang JX, Pan JM, Yin YJ, Wu RR, Dai XH, Dai JD, Gao L (2015) Thermo-responsive and magnetic molecularly imprinted Fe3O4@carbon nanospheres for selective adsorption and controlled release of 2,4,5-trichlorophenol. J Ind Eng Chem 25:321–328
51. Naldini L (2015) Gene therapy returns to centrestage. Nature 526:351–360
52. Cox DBT, Platt RJ, Zhang F (2015) Therapeutic genome editing: prospects and challenges. Nat Med 21:121–131
53. Morrison C (2018) Alnylam prepares to land first RNAi drug approval. Nat Rev Drug Discov 17:156–157
54. Wang F, Qin Z, Lu H, He S, Luo J, Jin C, Song X (2019) Clinical translation of gene medicine. J Gene Med 21:e3108
55. Ma CC, Wang ZL, Xu T, He ZY, Wei YQ (2020) The approved gene therapy drugs worldwide: from 1998 to 2019. Biotechnol Adv 40:107502–107525
56. Groen EJN, Talbot K, Gillingwater TH (2018) Advances in therapy for spinal muscular atrophy: promises and challenges. Nat Rev Neurol 14:214–245
57. Hoy SM (2019) Onasemnogene abeparvovec: first global approval. Drugs 79:1255–1262
58. Mendell JR, Pozsgai ER, Griffin DA, Mendell JR, Rodino-KlapacL R (2016) β-Sarcoglycan gene transfer decreases fibrosis and restores force in LGMD2E mice. Gene Ther 23:57–66
59. Khalil IA, Yamada Y, Harashima H (2018) Optimization of siRNA delivery to target sites: issues and future directions. Expert Opin Drug Deliv 15:1053–1065
60. Zhang X, Das SK, Passi SF, Uehara H, Bohner A, Chen M, Tiem M, Archer B, Ambati BK (2015) AAV2 delivery of Flt23k intraceptors inhibits murine choroidal neovascularization. Mol Ther 23:226–234
61. Li S, Ou M, Wang G, Tang LL (2016) Application of conditionally replicating adenoviruses in tumor early diagnosis technology, gene-radiation therapy and chemotherapy. Appl Microbiol Biotechnol 100:8325–8335
62. Singh SR, Grossniklaus HE, Kang SJ, Edelhauser HF, Ambati BK, Kompella UB (2009) Intravenous transferrin, RGD peptide and dual-targeted nanoparticles enhance anti-VEGF intraceptor gene delivery to laser-induced CNV. Gene Ther 16:645–659
63. Marano RJ, Toth I, Wimmer N, Brankov M, Rakoczy PE (2005) Dendrimer delivery of an anti-VEGF oligonucleotide into the eye: a long-term study into inhibition of laser-induced CNV, distribution, uptake and toxicity. Gene Ther 12:1544–1550
64. Kretzmann JA, Evans CW, Moses C, Sorolla A, Kretzmann AL, Wang E, Ho D, Hackett MJ, Dessauvagie BF, Smith NM, Redfern AD, Waryah C, Norret M, Iyer KS, Blancafort P (2019) Tumour suppression by targeted intravenous non-viral CRISPRa using dendritic polymers. Chem Sci 10:7718–7727
65. Kawakami S, Harada A, Sakanaka K, Nishida K, Nakamura J, Sakaeda T, Ichikawa N, Nakashima M, Sasaki H (2004) In vivo gene transfection via intravitreal injection of cationic liposome/plasmid DNA complexes in rabbits. Int J Pharm 278:255–262
66. Meng H, Mai WX, Zhang H, Xue M, Xia T, Lin S, Wang X, Zhao Y, Ji Z, Zink JI, Nel AE (2013) Codelivery of an optimal drug/siRNA combination using mesoporous silica nanoparticles too vercome drug resistance in breast cancer in vitro and in vivo. ACS Nano 7:994–1005
67. Guan XW, Guo ZP, Lin L, Tian HY, Chen SX (2016) Ultrasensitive pH triggered charge/size dual-rebound gene delivery system. Nano Lett 16:6823–6831
68. Zou Y, Zheng M, Yang W, Meng F, Miyata K, Kim HJ, Kataoka K, Zhong Z (2017) Virus-mimicking chimaeri cpolymer somesboost targeted cancer siRNA therapy *in vivo*. Adv Mater 9:1703272–1703285

69. Gupta N, Srivastava J, Singh LK, Singh AK, Shah K, Prasad R (2020) Epitope imprinting of outer membrane protein of neisseria meningitides. J Sci Res 64:220–228
70. Xu J, Merlier F, Avalle B, Vieillard V, Debré P, Haupt K, Tse Sum Bui B (2019) Molecularly imprinted polymer nanoparticles as potential synthetic antibodies for immunoprotection against HIV. ACS Appl Mater Interfaces 11:9824–9831
71. Muti M, Soysal M, Nacak FM, Gençdağ K, Karagözler AE (2015) A novel DNA probe based on molecularly imprinted polymer modified electrode for the electrochemical monitoring of DNA. Electroanalysis 27:1368–1377
72. Brahmbhatt H, Poma A, Pendergraff HM, Watts JK, Turner NW (2016) Improvement of DNA recognition through molecular imprinting: hybrid oligomer imprinted polymeric nanoparticles (oligoMIP NPs). Biomater Sci 4:281–287
73. Ren K, Banaei N, Zare RN (2013) Sorting inactivated cells using cell-imprinted polymer thin films. ACS Nano 7:6031–6036
74. You M, Yang S, Jiao F, Yang LZ, Zhang F, Hea PG (2016) Label-free electrochemical multi-sites recognition of G-rich DNA using multi-walled carbon nanotubes—supported molecularly imprinted polymer with guanine sites of DNA. Electrochim Acta 199:133–141
75. Ogiso M, Minoura N, Shinbo T, Shimizu T (2006) Detection of a specific DNA sequence by electrophoresis through a molecularly imprinted polymer. Biomaterials 27:4177–4182
76. Bartold K, Pietrzyk-Le A, Huynh TP, Iskierko Z, Sosnowska M, Noworyta K, Lisowski W, Sannicolò F, Cauteruccio S, Licandro E, D'Souza F, Kutner W (2017) Programmed transfer of sequence information into a molecularly imprinted polymer for hexakis (2,2′-bithien-5-yl) DNA analogue formation toward single-nucleotide-polymorphism detection. ACS Appl Mater Interfaces 9:3948–3958
77. Mura S, Nicolas J, Couvreur P (2013) Stimuli-responsive nanocarriers for drug delivery. Nat Mater 12:991–1003
78. Qin L, He XW, Zhang W, Li WY, Zhang YK (2009) Macroporous thermosensitive imprinted hydrogel for recognition of protein by metal coordinate interaction. Anal Chem 81:7206–7216
79. Parisi OI, Morelli C, Puoci F, Saturnino C, Caruso A, Sisci D (2014) Magnetic molecularly imprinted polymers (MMIPs) for carbazole derivative release in targeted cancer therapy. J Mater Chem B 2:6619–6625
80. Xu SF, Li JH, Song XL, Liu JS, Chen LX (2013) Photonic and magnetic dual responsive molecularly imprinted polymers: preparation, recognition characteristics and properties as a novel sorbent for caffeine in complicated samples. Anal Meth 5:124–133
81. Griffiths JR (1991) Are cancer cells acidic? Br J Cancer 64:425–427
82. Chianella I, Guerreiro A, Moczko E, Caygill JS, Piletska EV, De VS (2013) Direct replacement of antibodies with molecularly imprinted polymer (MIP) nanoparticles in ELISA—development of a novel assay for vancomycin. Anal Chem 85:8462–8468
83. Wang YJ, Zhao XY, Chen Q, Zhai HJ, Li SD (2015) B11: a moving subnanoscale tank tread. Nanoscale 7:16054–16060
84. Hashemi-Moghaddam H, Zavareh S, Karimpour S, Madanchi H (2017) Evaluation of molecularly imprinted polymer based on HER2 epitope for targeted drug delivery in ovarian cancer mouse model. React Funct Polym 121:82–90
85. Liu S, Bi QY, Long YY, Li ZX, Bhattacharyya S, Li C (2017) Inducible epitope imprinting: 'generating' the required binding site in membrane receptors for targeted drug delivery. Nanoscale 9:5394–5397
86. Edwardraja S, Lee SG (2010) *In vivo* based N-terminal specific addition of bio-orthogonal reactive group on single chain Fv fragment. In: Fall KSBB meeting & international symposium, vol 11, pp 498–501
87. Stimmel JB, Merrill BM, Kuyper LF, Moxham CP, Hutchins JT, Fling ME, Kull FC (2000) Site-specific conjugation on serine right-arrow cysteine variant monoclonal antibodies. J Biol Chem 275:30445–30450

Chapter 10
Outlook

Xuemei Wang, Pengfei Huang, and Zheng Zhou

Molecularly imprinted polymers (MIPs) can specifically recognize target molecules due to their special imprinted cavity structure. It exhibits many excellent properties such as high selectivity, stability, and durability against harsh conditions. As shown in Fig. 10.1, many different target molecules have been studied and published in the last 2 years [1]. The application field of molecular imprinting is mainly oriented to separation, purification, and quantification of different compounds, and it has great potential in manufacturing satisfactory pharmaceutical dosage forms. This chapter aims to review new strategies for developing different types of MIP for drug delivery systems (DDS). Although its application in DDS is still in its infancy, the application of MIP design and devices in closely related fields (such as diagnostic sensors) have attracted more and more attention [2].

Drug delivery is the method or process of administering a pharmaceutical compound to achieve an optimal therapeutic effect in humans or animals. A Drug Delivery System (DDS) can help improve drug absorption, extend the effective lifetime of a drug in the body, and deliver the drug specifically to the affected area. A DDS can thus maximize the effectiveness of drugs. DDS conducted extensive research and achieved rapid growth over the past few decades. The delivery of a drug to its target plays a crucial role in the effectiveness of drug delivery strategies. Some drugs require a certain concentration to exert the best therapeutic effect; on the contrary, the higher or lower content of some drugs may have toxic or negative effects on the target. The time required to release the drug at a controlled rate and precise target, biocompatibility or biodegradability, so that the delivery system is converted to be harmlessly eliminated from the body. The release of non-toxic fragments is a few problems in DDS. Therefore, an ideal delivery vehicle will ensure

X. Wang (✉) · P. Huang · Z. Zhou
Key Laboratory of Bioelectrochemistry & Environmental Analysis of Gansu Province, College of Chemistry and Chemical Engineering, Northwest Normal University, Lanzhou, People's Republic of China

Z. Liu et al. (eds.), *Molecularly Imprinted Polymers as Advanced Drug Delivery Systems*, https://doi.org/10.1007/978-981-16-0227-6_10

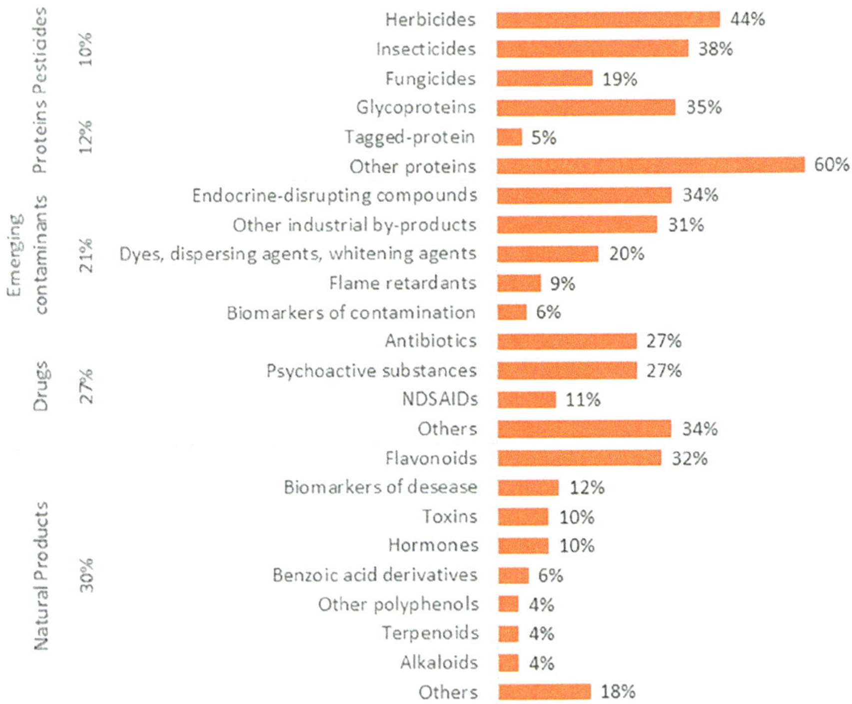

Fig. 10.1 Groups of molecules for which MIPs have been recently developed. *NSAIDs* nonsteroidal anti-inflammatory drugs. (Reproduced with permission from ACS Publications; Pichon et al. [1])

that the drug is released in the correct location, in the correct dosage, and in the required time, and has no toxic effect on the host.

To minimize the problem, an interdisciplinary approach that combines polymer chemistry, molecular biology, and pharmaceutical sciences is constantly being adopted. Due to its universal performance and easy-to-design architecture, polymer chemistry has always been at the forefront of controlling DSS. Therefore, it is reported that most of the currently controlled DDSs are based on polymers. Mechanisms that control the rate of drug release from the majority of current polymeric drug delivery systems can be characterized as being controlled by diffusion, swelling systems, erosion [3, 4], or by an external stimulus [5, 6], Fig. 10.2 is the four traditional categories of drug delivery systems based on mechanism of drug release [7]. In this chapter, the future prospects will be summarized in three areas including improved formats for polymerization, enhancing site accessibility, and post-polymerization modification of polymers.

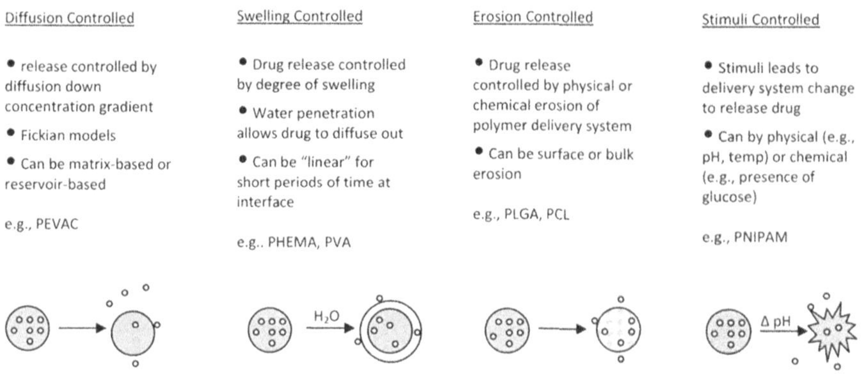

Fig. 10.2 Summary of four traditional categories of drug delivery systems based on mechanism of drug release—diffusion-controlled, swelling-controlled, erosion-controlled, and stimuli-controlled systems. (Reproduced with permission from John Wiley & Sons; Wang and Von Recum [7])

10.1 Future Prospects-Improved Formats for Polymerization

At present, MIPs has several preparation methods according to the different performance requirements, such as bulk polymerization, in suit polymerization, seed swelling polymerization, suspension polymerization, surface molecular imprinting polymerization, hollow molecular imprinting polymerization, etc. Among them, seed swelling, suspension, surface imprinting, and hollow molecular imprinting are more commonly used and improved polymerization formats, especially in the DDS field. In this section, several novel kinds of polymerization formats will be summarized detailed. They are emerging and promising polymerization formats in DDS, such as hollow molecular imprinting technology, ring-opening metathesis polymerization (ROMP), Janus molecularly imprinted technology.

Hollow molecular imprinting, also known as the sacrificial carrier method, is a synthetic imprinting polymer method based on surface molecular imprinting [8, 9]. After the surface molecular imprinting is synthesized, the carrier nucleus in the surface molecular imprinting is removed by chemical dissolution or erosion to obtain the molecularly imprinted polymer with a hollow structure. The imprinted polymers synthesized by this method not only have the advantages of surface molecularly imprinted polymers but also have a larger adsorption capacity and a faster mass transfer rate than surface molecularly imprinted polymers. Erosion control systems have gained popularity with the development of biodegradable polymers, which use chemical or physical polymers or material loss to control drug delivery [10–12]. Releasing drugs from these systems can be complicated. However, erosion is usually a combination of material transport and chemical reactions, which may involve drug dissolution, polymer degradation, porosity generation, microenvironmental changes in pH, diffusion in the polymer matrix, and autocatalysis [4]. Release from erosion-controlled systems can be bi- or multi-phasic

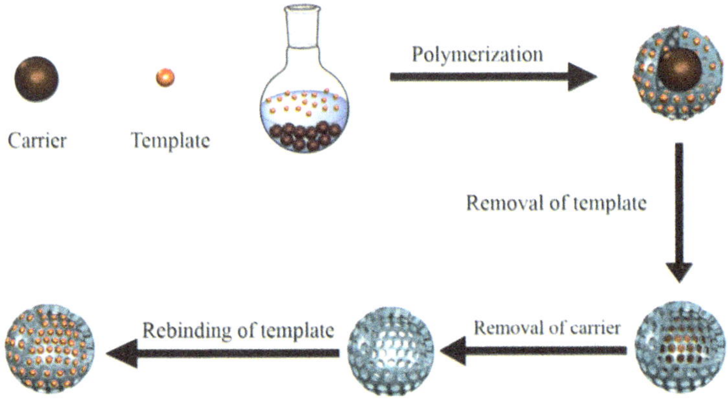

Fig. 10.3 The scheme of hollow molecularly imprinted polymers

depending on the level of polymer degradation in the system. Figure 10.3 is the scheme of hollow molecularly imprinted polymers.

There is very little work on ring-opening metathesis polymerization (ROMP) as the MIP matrix [13]. The latest discovery of well-defined transition metal catalysts for the metathesis polymerization of olefin compounds provides a wide range of unique materials [14–18]. The most effective catalysts are those derived from ruthenium; these have been developed and promulgated by Grubbs. The ROMP matrices using Grubbs catalyst provides an insoluble cross-linked polymer. For example, we can find the difference between radical polymerization and ROMP matrices in molecular imprinting technology from Figs. 10.4 and 10.5 [19].

Because of the asymmetric dual functions, Janus particles have been widely used as solid stabilizers, imaging probes, targeting sensors, interfacial catalysts, and photonic materials in various fields [20]. In general, a variety of techniques can be used to prepare Janus particles. Recently, methods for controlling the structural diversity of Janus particles mainly include block copolymer self-assembly methods, phase separation methods, competitive adsorption methods, integration of microfluidic devices, and photolithographic polymerization or photopolymerization [21–24]. The control of the anisotropic function of Janus particles can be successfully achieved using the above strategy. However, selective loading of chemical or biological components (such as dyes, probes, drugs, and image contrast agents) on Janus particles is still difficult to achieve [25]. This defect hinders the practical application of Janus particles. Therefore, the selective loading/unloading of target reagents onto Janus particles is a major challenge. So far, there have been studies on the construction of Janus particles with specific molecular recognition capabilities. For example, monodisperse MIP particles containing amino groups (MIP-NH_2 particles) were synthesized via a two-step precipitation polymerization reaction (Fig. 10.6) [26] following a similar procedure described by Hajizadeh et al. [27].

Besides, with the development of materials chemistry from macro-systems to micro- and nano-systems, the application range of DDS has also shrunk to

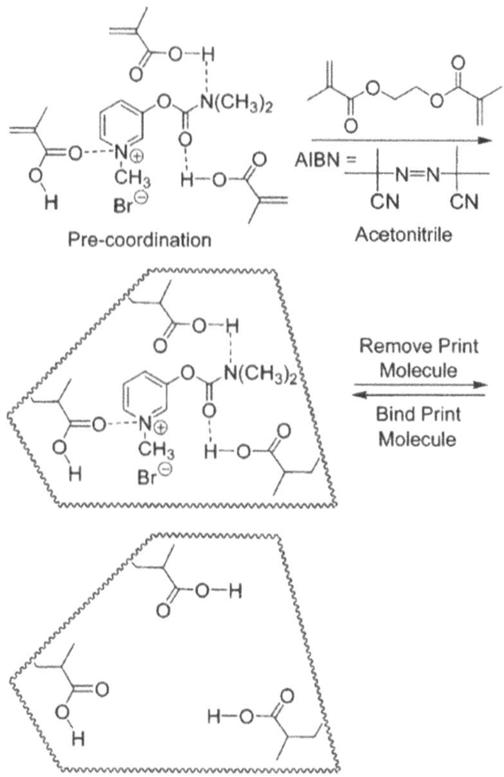

Fig. 10.4 Schematic draw of the imprinting process of ring-opening metathesis polymerization by choosing acetylcholine binding agents. (Reproduced with permission from ACS Publications; Enholm et al. [19])

nano-materials, and DDS has also expanded from small drugs to large biomolecules, such as proteins and peptides. Therefore, numerous reports aiming to cater to the needs of polymeric nano-sized DD materials that could accommodate small and larger drug molecules facilitating the easy release of intended drugs have been adapted and employed [28–30]. Also, a review discussing a new type of microelectrochemical system or MEMS-based DDS called microchip (implantable microchip) has recently been published. This report presented an overview of the investigations on the feasibility and application of microchip as an advanced DDS, and it's commercial manufacturing materials and methods [31].

10.2 Future Prospects-Enhancing Site Accessibility

MIPs has played an important role in fabricating predefined drug selectivity in synthesized polymer products, which provides for significant changes in physico-chemical property and recognition of its intended application. The physical characteristics of the cross-linked material that is imprinted to create a shaped imprint cavity have three-dimensional interaction sites that contribute to the chemical

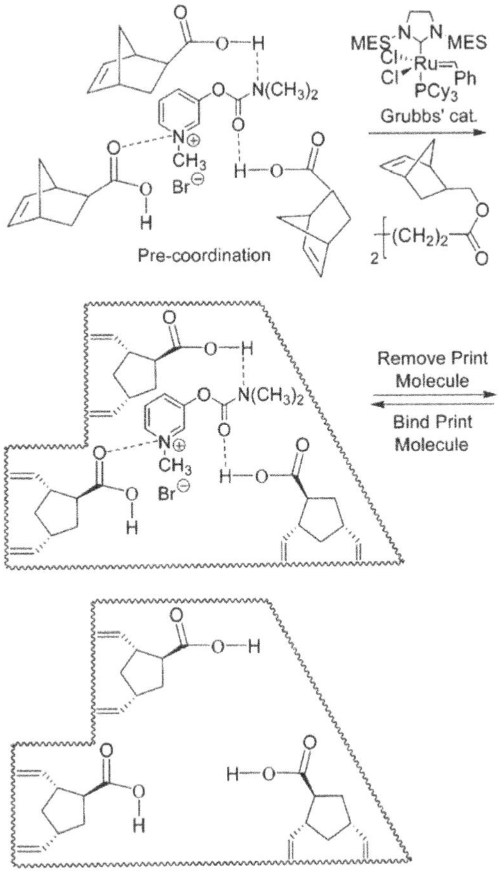

Fig. 10.5 Schematic draw of the imprinting process of radical polymerization by choosing acetylcholine binding agents. (Reproduced with permission from ACS Publications; Enholm et al. [19])

properties, which is useful for DDS. A novel approach to design and creation of a recognition cavity is especially important, to implement the functionality of necessary stereochemical feature in artificial materials, and the discriminatory capacity of the specific sites for the pharmaceutically active agents [32]. This technology uses molecular templates (printed molecules) to create recognition sites in the polymer matrix during synthesis. In other words, the development of DDS may have the potential to enhance site accessibility. In DDS, release is a response to changes in environmental conditions, which directly affects the binding of drugs (competitive binding) (Fig.10.7) [33] or the hydrolysis of conjugates (Fig. 10.8) [32] or changes in polymer swelling state (volume phase transition induced by an external stimulus) (Fig. 10.9) [2].

An activation-modulated delivery may be achieved with an imprinted gel that releases the drug because of the competitive binding to the polymer of another substance in solution [2]. The network includes non-imprinted drugs, and when imprinted molecules are present in a specific medium, the network will bind to it and release the drug. The network will stop releasing the imprinting substance when the

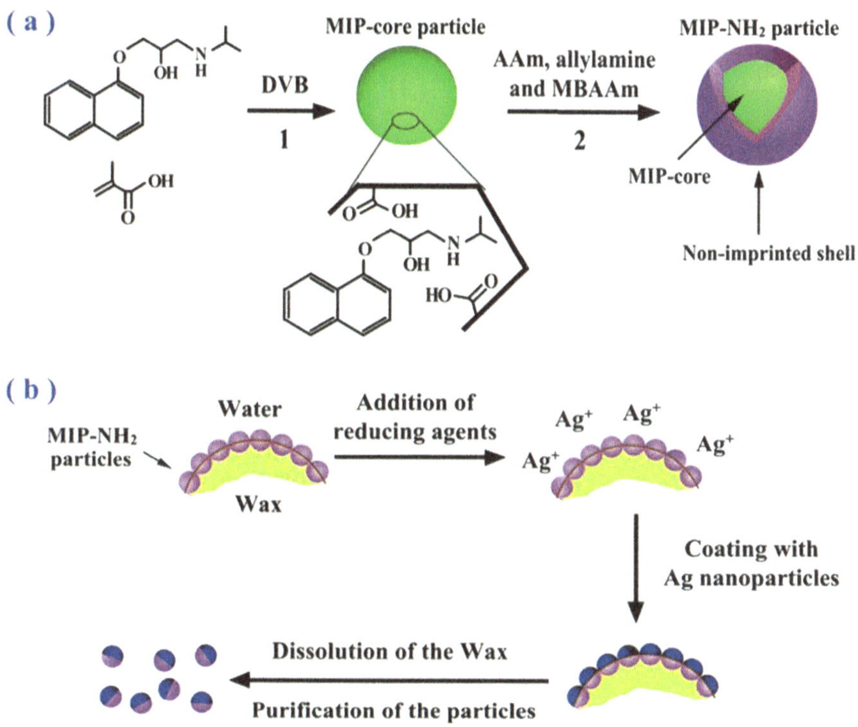

Fig. 10.6 (**a**) Schematic diagram of the synthesis of monodisperse MIP-NH2 microspheres. Step 1: Cross-link polymerization to generate propranolol-imprinted sites. Step 2: Copolymerization and grafting of amino groups. (**b**) Stepwise synthesis of Janus colloidal particles. Schematic representation of the synthesis of monodisperse MIP-NH_2 microspheres. Step 1: generation of propranolol-imprinted sites by cross-linking polymerization. Step 2: grafting amine groups by copolymerization. (**b**) Step-by-step fabrication of Janus colloidal particles. (Reproduced with permission from Royal Society of Chemistry Publications; Huang and Shen [26])

concentration of the imprinting substance decreases. This competitive binding indicates that the non-imprinted species bound to the imprinted polymer particles can be replaced by template molecules. Through this, it can be confirmed that the recognition site is very important for the DDS capacity and improving the mass transfer efficiency.

A particularly useful approach to modulate drug delivery consists of creating erosion systems in order to enhance active sites from which the drug cannot be released unless the polymer degrades or polymer-drug bonds are broken. The external conditions that can induce these processes are, usually, extreme physiological pH or the catalytic activity of certain enzymes [34]. To enhance the hydrolysis of polymer-drug ester or amide bonds under mild pH conditions, Karmalkar et al. proposed incorporating imidazole groups (nucleophilic catalyst) near the drug linkage using a molecular imprinting technique [35].

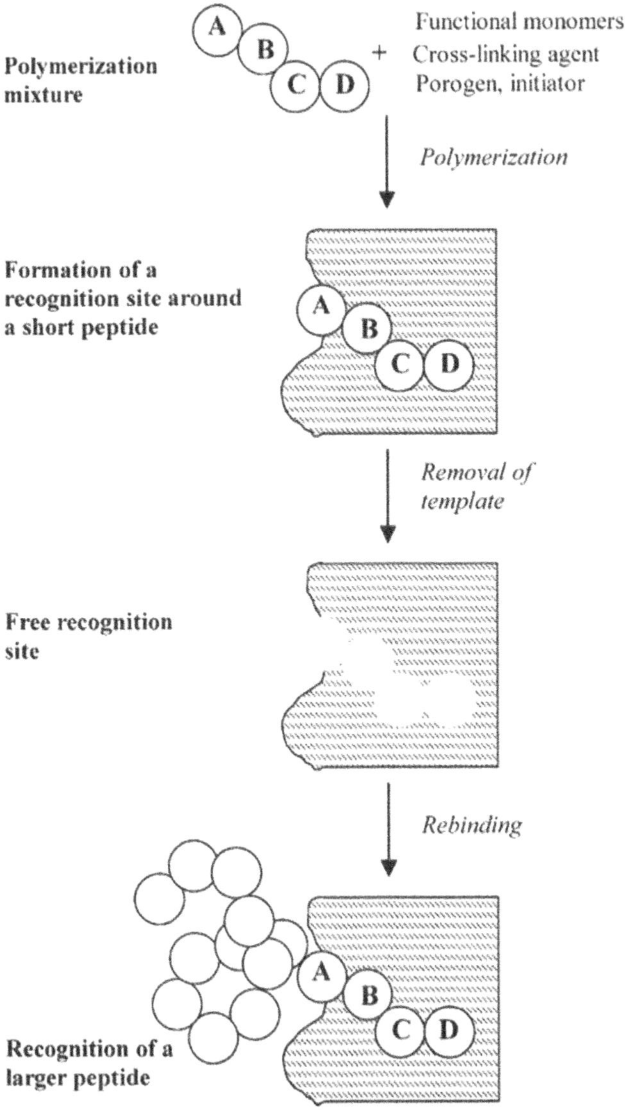

Fig. 10.7 Schematic draw of the imprinting process of a peptide using the epitope approach. (Reproduced with permission from Elsevier Publications; Rachkov and Minoura [33])

Polymer gels that modify their structure and, in consequence, their properties in response to changes in the physicochemical characteristics of the physiological medium are very promising candidates to achieve optimum control of the moment and rate of drug release [36, 37]. The sensitivity of the recognition site and the combination of imprinting have considerable practical advantages: imprinting has a high load capacity for specific molecules, and the ability to respond to external

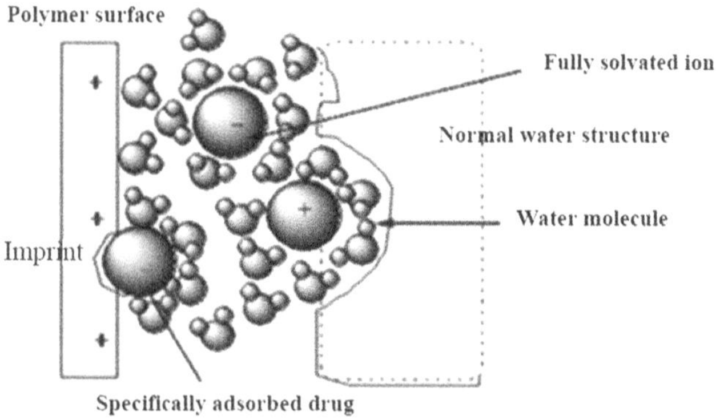

Fig. 10.8 A model of the solid/liquid interface with specific adsorption on the imprinted layer and non-specific adsorption of the substrate. (Reproduced with permission from Longdom Publishing; Suedee [32, 45])

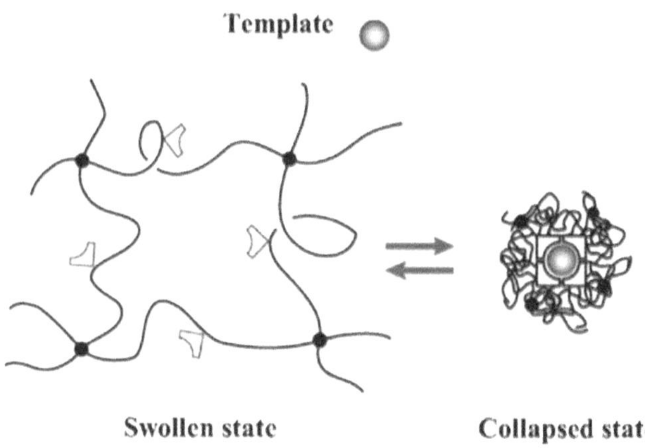

Fig. 10.9 Diagram of the recognition process of a template by a stimuli sensitive imprinted hydrogel. (Reproduced with permission from Elsevier Publications; Alvarez-Lorenzo and Concheiro [2])

recognition sites is conducive to improving the affinity of the network to the target molecule, thereby providing a regulatory effect or loading/releasing capability handover process [38]. Theoretically, the polymer network has the ability to remember a specific conformation after a sharp change in the degree of swelling.

In addition, by combining the asymmetric structure of Janus particles with the specific molecular recognition ability of MIPs, Janus MIP particles synthesized by waxy Pickering emulsion can be used as self-propelled transporters for controlled drug delivery [26]. Considering their use as DDS, due to the presence of recognition sites from MIP, when administered orally, MIP has great potential to target drugs to specific areas of the gastrointestinal tract.

10.3 Future Prospects-Post-polymerisation Modification of Polymers

It has been a well-known fact that polymeric materials are one of the very exciting and useful drug delivery vehicles. The drugs are dispersed within the polymer matrix designed to release them over a prolonged period or under certain physiological conditions; therefore, various harmful effects caused by a narrow therapeutic window and the concentration lower than the required level are overcome. However, these polymer matrices have been found to have a burst drug release, which has potentially negative consequences for patients [39–41]. Due to their outstanding and unique functions in the previous section, MIP can be used as a potential novel drug delivery carrier.

Generally speaking, MIP matrices rely on a high degree of internal cross-linking to retain complementary cavities and use these cavities for drug depots. However, preparing molecularly imprinted polymer gels that are less tightly cross-linked is advantageous for drug delivery, especially when water-soluble monomers are used. These types of MIPs will show expected changes on their surface with environmental changes (including changes in solvent quality, temperature, and ions). The drugs are dispersed within the polymer matrix designed to release them over a prolonged period or under certain physiological conditions; thus, overcoming the various harmful effects caused by a narrow therapeutic window and the concentration below required levels. The improved polymeric matrixes have been developed, but their practical application were limited by the feedback-controlled release and other problems with many polymeric DDS [41].

Recently, the post-polymerization modification (PPM) method is used as a vehicle for preparing functional polymers. PPM is a synthesis method that quantitatively converts the inert group in a polymer into other functional groups. PPM is a synthetic method in which the inert functional groups are quantitatively transformed into other functional groups by direct polymerization or copolymerization of the chemically inert monomers under polymerization conditions in the following steps [42]. It is usually carried out under mild conditions, and the transformed functional groups have some tolerance. In this part, the evolution of eight main classes of post-polymerization modification reactions will be outlined (Fig. 10.10).

The usual method used to prepare functional polymers is to properly protect the functional monomers and then remove the functional groups to achieve the purpose of preparation. Direct polymerization of functional monomers is a good chemical strategy, but only a few functional monomers are suitable for direct polymerization, and this method will affect the main chain structure of polymers to a certain extent. If the prepared polymer can be modified (i.e., PPM method) to prepare functional polymers, the above problems will be readily solved.

Besides, the imprinted hydrogels modulate (collapse and swell) to suit the conditions where the polymer is going to be used to release its load of the therapeutic agent. According to reports, compared with other controlled polymer systems, the drug release profile of MIP-based DDS devices is not impressive, and this can be

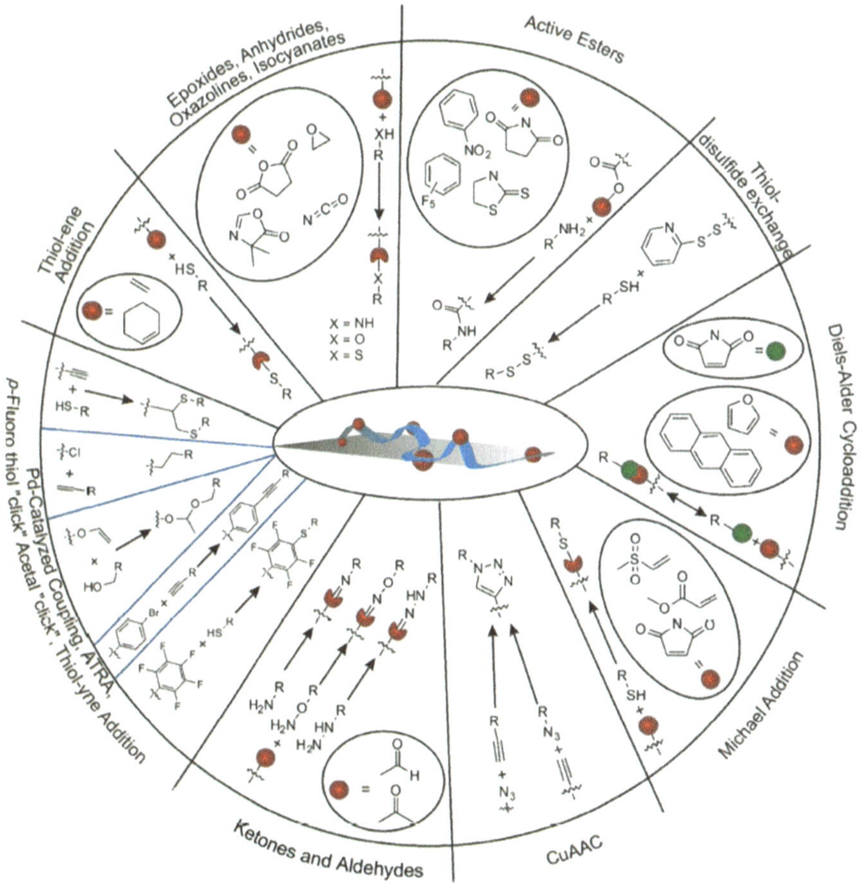

Fig. 10.10 Schematic illustration of the main classes of reactions that can be used for the preparation of functionalized polymers via post-polymerization modification. (Reproduced with permission from John Wiley & Sons; Günay et al. [42])

achieved with slight modifications. However, they appear as intelligent feedback-controlled DDS, which can perceive the surrounding environment and trigger drug release when any disease occurs overexpression of biomarkers [43, 44].

One of the many excellent advantages of MIPs is that they display high stability and durability against harsh conditions (e.g., thermal, mechanical, and highly acidic and basic pH conditions). MIPs can be stored in the dry state at ambient temperature for several years without losing their recognition features, resistance to crushing and grinding and withstand in highly acidic and basic conditions. These features make MIPs suitable candidates for sustained drug delivery formulations as they show high stability in the human body conditions, particularly gastrointestinal conditions, where pH is highly acidic and non-polymeric formulations are shown to be highly unstable leading to burst release. Another advantage of using MIPs as DDS is that the covalent or non-covalent preparations of MIPs may regulate the drug release (i.e., controlled release) actions by increasing and decreasing the residence time of drug within the polymer, respectively, to avoid its side effect due to overconcentration of the drug within the body at a particular time [32, 45]. Therefore, due to the affinity of the template to the functional monomer, MIP can help achieve the sustained release of the drug, thereby prolonging the residence time of the drug in the human body.

10.4 Conclusion

Reviewing numerous review articles and reports, the MIP-based DDS system has greater potential and is an emerging field, representing one of the main research and development areas in the field of drug delivery today. Reviewing numerous reports and review articles, it is worth noticing that MIPs-based DDS systems exhibit greater potential and a burgeoning field representing one of the major research and development focus areas in drug delivery today. However, it also faces obvious obstacles, such as controlling balance pharmacokinetics and pharmacodynamics, toxicity and biocompatibility of employed polymer, selective recognition, the efficacy of drug loading and release, behavior of MIPs in the surrounding environment. For example, the demand for biocompatibility of MIPs significantly reduced the choice of effective functional monomers and cross-linker for the imprinting process, because many of them are highly toxic. Second, most of the polymerization reactions for the synthesis of MIP are carried out in organic solvents, which is contrary to the aqueous environment required for in-vitro mimicking. Therefore, the optimization of experimental parameters is indispensable. In order to improve the efficacy of MIP to deliver drugs, evenly distributed binding sites and morphologies are required, and therefore, better polymerization methods need to be utilized. Furthermore, the reproducibility of the synthesis is also vital to ensure the robustness and practicality of the synthesized MIP.

Currently, MIP as a drug delivery device has not been found to be put into any commercial application. A lot of research is still needed to overcome these challenges. As mentioned earlier, although some applications of MIPs have been developed, the development of the molecular imprinting approach for DDS is still in its infancy. However, it is foreseeable that in the next few years, with the progress of other fields, this technology will also make major breakthroughs in the future. In the evolutionary route to improve the applicability of drug delivery imprinting, the focus should be on the rational design of imprinting systems and the application of molecular imprinting in water. To optimize the nature and quantity of functional monomers, the specificity and affinity of the template should be improved, and the time and material consumption methods of "trial and error" should be overcome [46]. This can be achieved through the use of isothermal titration calorimetry, rapid experimental screening procedures [47], and combinatorial libraries [48], which have proven to be suitable methods for studying the thermodynamics of evaluating the efficiency of molecular imprinting processes and molecular recognition. In the future, the development direction of DDS molecular imprinting technology is as follows:

1. The concept of intelligent drug delivery which refers to the predictable release of a drug in response to specific stimuli, such as the presence of another molecule has been a well-established drug release way from MIP carrier. Some excellent reviews on the investigation of various aspects of MIPs in drug delivery have been published [7, 49–53]. In addition to numerous research reports published on various concepts of MIP-based DDS, some important contributions in the form of book chapters have also been devoted to the extensive introduction and applications of MIP-based intelligent drug delivery [54].
2. Transporter-targeted nano-DDS is considered as a promising nano-platform that can achieve efficient drug delivery. The basic strategy is to modify the nano-DDS with specific substrates of transporters, including derivatives and natural substrates (such as amino acids, carnitine, glucose, vitamins, and choline).
3. Magnetic drug targeting play a key role in capture or guide drugs to designated sites [54]. A suitable magnet system is the prerequisite to achieve successful magnetic drug targeting. At present, although there are many different types of magnet systems, their practical applications are often limited to some simple cases, such as diseases on the skin surface or close to the body surface. Due to the above reasons, the scope of application and promotion of targeted magnetic drug delivery is limited. For the interior of the human body, the research on the 3D precise targeting magnet system is still in the basic research stage. Many models are preliminary and have not been fully tested. 3D localized high magnetic field may be the best way to achieve precise targeting, but there is no real 3D precise targeting yet. Therefore, more fundamental researches are still indispensable to realize the wide application of magnetic drug targeting.
4. Post-imprinting modification (PIM) is an innovative strategy for generating MIPs analogous to biosynthetic proteins. New functionalities are introduced, in a site-directed manner, into a molecularly imprinted cavity. The monomer groups in the cavity can be chemically modified to introduce new functions, such as fluorescent signals, light responsiveness, and fine-tuned binding characteristics, and on/off switching of binding activity [55].

Now, MIP has been used to analyze potential methods such as sample enrichment, chromatographic separation, molecular recognition, and electrochemical sensors, but as described in this article, the practical application of MIP as an active biomedical device is still in the early stages of development. Although MIP-based drug carriers have been used in multiple drug delivery routes, the drug release efficacy and biocompatibility of MIP are still huge challenges that must be resolved.

However, it can be expected that based on current trends and procedures, it is most likely that some of the more exciting future developments in drug delivery and therapeutic monitoring will rely on some form of real-time analysis to affect intelligence, based on the feedback response of MIP-based DDS systems result. It must be proven soon that the MIP-based DDS system may be a real competitor of the existing drug delivery carrier materials in the field of biomedical equipment. In addition, some elegant studies have established the potential for controlled drug delivery based on implantable microchips. The hybrid of MIP and microchip-based DDS may prove to be a promising biomedical device in the near future.

References

1. Pichon V, Delaunay N, Combes A (2020) Sample preparation using molecularly imprinted polymers. Anal Chem 92(1):16–33
2. Alvarez-Lorenzo C, Concheiro A (2004) Molecularly imprinted polymers for drug delivery. J Chromatogr B 804(1):231–245
3. Lin CC, Metters AT (2006) Hydrogels in controlled release formulations: network design and mathematical modeling. Adv Drug Deliv Rev 58(12–13):1379–1408
4. Arifin DY, Lee LY, Wang C-H (2006) Mathematical modeling and simulation of drug release from microspheres: implications to drug delivery systems. Adv Drug Deliv Rev 58 (12–13):1274–1325
5. Oupicky D, Bisht HS, Manickam DS, Zhou QH (2005) Stimulus-controlled delivery of drugs and genes. Expert Opin Drug Deliv 2(4):653–665
6. Alarcón CDLH, Pennadam S, Alexander C (2005) Stimuli responsive polymers for biomedical applications. ChemInform 34(3):276–285
7. Wang NX, Recum HAV (2011) Affinity-based drug delivery. Macromol Biosci 11(3):321–332
8. Li H, Li Y, Li Z, Peng X, Li Y, Li G, Tan X, Chen G (2012) Preparation and adsorption behavior of berberine hydrochloride imprinted polymers by using silica gel as sacrificed support material. Appl Surf Sci 258(10):4314–4321
9. Pan J, Zou X, Yan Y, Wang X, Guan W, Han J, Wu X (2010) An ion-imprinted polymer based on palygorskite as a sacrificial support for selective removal of strontium(II). Appl Clay Sci 50 (2):260–265
10. Siepmann J, Gopferich A (2001) Mathematical modeling of bioerodible, polymeric drug delivery systems. Adv Drug Deliv Rev 48(2–3):229–247
11. Kimura H, Ogura Y (2001) Biodegradable polymers for ocular drug delivery. Ophthalmologica 215(3):143–155
12. Das S, Rao GHR, Robert F (2000) Gladwin, colchicine encapsulation within poly(ethylene glycol)-coated poly(lactic acid)/poly(epsilon-caprolactone) microspheres-controlled release studies. Drug Deliv 7(3):129–138
13. Patel A, Fouace S, Steinke JHG (2002) Enantioselective molecularly imprinted polymers via ring-opening metathesis polymerisation. Chem Commun 1:88–89
14. Sundararajan G (1994) Metathetical polymerization—a review. J Sci Ind Res 53(6):418–432
15. Yun J, Marinez ER, Grubbs RH (2004) A new ruthenium-based olefin metathesis catalyst coordinated with 1, 3-dimesityl-1, 4, 5, 6-tetrahydropyrimidin-2-ylidene: synthesis, X-ray structure, and reactivity. Organometallics 23(18):4172–4173
16. Love JA, Sanford MS, Day MW, Grubbs RH (2003) Synthesis, structure, and activity of enhanced initiators for olefin metathesis. J Am Chem Soc 125(33):10103–10109
17. Trnka TM, Morgan JP, Sanford MS, Wilhelm TE, Scholl M, Choi T-L, Ding S, Day MW, Grubbs RH (2003) Synthesis and activity of ruthenium alkylidene complexes coordinated with phosphine and N-heterocyclic carbene ligands. J Am Chem Soc 125(9):2546–2558
18. Feast W (1992) Applications of ROMP in the synthesis of new materials. In: Makromolekulare Chemie. Macromolecular symposia, Wiley Online Library, pp 317–326
19. Enholm EJ, Allais F, Martin RT, Mohamed R (2006) A comparison of a radical polymerization vs ROMP matrix for molecular imprinting. Macromolecules 39(23):7859–7862

20. Walther A, Muller AH (2013) Janus particles: synthesis, self-assembly, physical properties, and applications. Chem Rev 113(7):5194–5261
21. Dendukuri D, Doyle PS (2009) The synthesis and assembly of polymeric microparticles using microfluidics. Adv Mater 21(41):4071–4086
22. Lattuada M, Hatton TA (2011) Synthesis, properties and applications of Janus nanoparticles. Nano Today 6(3):286–308
23. Perro A, Reculusa S, Ravaine S, Bourgeat-Lami E, Duguet E (2005) Design and synthesis of Janus micro-and nanoparticles. J Mater Chem 15(35–36):3745–3760
24. Yu Z, Wang CF, Ling L, Chen L, Chen S (2012) Triphase microfluidic-directed self-assembly: anisotropic colloidal photonic crystal supraparticles and multicolor patterns made easy. Angew Chem 124(10):2425–2428
25. Jung CY, Kim JS, Kim HS, Ha JM, Kim ST, Lim HJ, Koo SM (2012) Selective surface reactions for Janus ORMOSIL particles with multiple functional groups using an ordered monolayer film at liquid–liquid interface. J Colloid Interface Sci 367(1):257–263
26. Huang C, Shen X (2014) Janus molecularly imprinted polymer particles. Chem Commun 50 (20):2646–2649
27. Hajizadeh S, Xu C, Kirsebom H, Ye L, Mattiasson B (2013) Cryogelation of molecularly imprinted nanoparticles: a macroporous structure as affinity chromatography column for removal of β-blockers from complex samples. J Chromatogr A 1274:6–12
28. Gaspar RS, Duncan R (2009) Polymeric carriers: preclinical safety and the regulatory implications for design and development of polymer therapeutics. Adv Drug Deliv Rev 61 (13):1220–1231
29. De Souza R, Zahedi P, Allen CJ, Piquette-Miller M (2010) Polymeric drug delivery systems for localized cancer chemotherapy. Drug Deliv 17(6):365–375
30. Demetzos C, Pippa N (2014) Advanced drug delivery nanosystems (aDDnSs): a mini-review. Drug Deliv 21(4):250–257
31. Sutradhar KB, Sumi CD (2014) Implantable microchip: the futuristic controlled drug delivery system. Drug Deliv 23(1):1–11
32. Suedee R (2013) The use of molecularly imprinted polymers for dermal drug delivery. Pharm Anal Acta 4(8):1–23
33. Rachkov A, Minoura N (2001) Towards molecularly imprinted polymers selective to peptides and proteins. The epitope approach. Biochim Biophys Acta Protein Struct Mol Enzymol 1544 (1–2):255–266
34. Shah S, Kulkarni M, Mashelkar R (1990) Release kinetics of pendant substituted bioactive molecules from swellable hydrogels: role of chemical reaction and diffusive transport. J Membr Sci 51(1–2):83–104
35. Karmalkar RN, Kulkarni MG, Mashelkar RA (1997) Pendent chain linked delivery systems: II. Facile hydrolysis through molecular imprinting effects. J Control Release 43(2–3):235–243
36. Yuk SH, Bae YH (1999) Phase-transition polymers for drug delivery. Crit Rev Ther Drug Carrier Syst 16(4):385–423
37. Okano T, Bae YH, Jacobs H, Kim SWJ (1990) Thermally on-off switching polymers for drug permeation and release. J Control Release 11(1–3):255–265
38. Watanabe M, Akahoshi T, Tabata Y, Nakayama D (1998) Molecular specific swelling change of hydrogels in accordance with the concentration of guest molecules. J Am Chem Soc 120 (22):855–861
39. Kumari A, Yadav SK, Yadav SC (2010) Biodegradable polymeric nanoparticles based drug delivery systems. Colloids Surf B Biointerfaces 75(1):1–18
40. Mora-Huertas CE, Fessi H, Elaissari A (2010) Polymer-based nanocapsules for drug delivery. Int J Pharm 385(1–2):113–142
41. Vilar G, Tulla-Puche J, Albericio F (2012) Polymers and drug delivery systems. Curr Drug Deliv 9(4):367–394
42. Günay KA, Theato P, Klok HA (2013) Standing on the shoulders of Hermann Staudinger: post-polymerization modification from past to present. J Polym Sci Part A Polym Chem 51(1):1–28

43. Cunliffe D, Kirby A, Alexander C (2005) Molecularly imprinted drug delivery systems. Adv Drug Deliv Rev 57(12):1836–1853
44. Sellergren B, Allender CJ (2005) Molecularly imprinted polymers: a bridge to advanced drug delivery. Adv Drug Deliv Rev 57(12):1733–1741
45. Suedee R (2013) Novel strategic innovations for designing drug delivery system using molecularly imprinted micro/nanobeads. Int J Pharm Sci Rev Res 20(2):235–268
46. Nicholls IA, Adbo K, Andersson HS, Andersson PO, Wikman S (2001) Can we rationally design molecularly imprinted polymers? Anal Chim Acta 435(1):9–18
47. Lanza F, Sellergren B (1999) Method for synthesis and screening of large groups of molecularly imprinted polymers. Anal Chem 71(11):2092–2096
48. Takeuchi T, Fukuma D, Matsui J (1999) Combinatorial molecular imprinting: an approach to synthetic polymer receptors. Anal Chem 71(2):285–290
49. Byrne ME, Park K, Peppas NA (2002) Molecular imprinting within hydrogels. Adv Drug Deliv Rev 54(1):149–161
50. Hillberg AL, Brain KR, Allender CJ (2005) Molecular imprinted polymer sensors: implications for therapeutics. Adv Drug Deliv Rev 57(12):1875–1889
51. Kryscio DR, Peppas NA (2010) Mimicking biological delivery through feedback-controlled drug release systems based on molecular imprinting. AIChE J 55(6):1311–1324
52. Rangasamy M, Kugalur GP (2010) Recent advances in novel drug delivery systems. Int J Res Ayurv Pharm 1(2):2025–2027
53. Luliński P (2013) Molecularly imprinted polymers as the future drug delivery devices. Acta Pol Pharm 70(4):601–609
54. Alvarez-Lorenzo C, Concheiro A (2013) From drug dosage forms to intelligent drug-delivery systems: a change of paradigm. Smart Mater Drug Deliv 1(1):1–32
55. Takeuchi T, Sunayama H (2018) Beyond natural antibodies—a new generation of synthetic antibodies created by post-imprinting modification of molecularly imprinted polymers. Chem Commun 54(49):6243–6251

The manufacturer's authorised representative in the EU is Springer Nature Customer Service Centre GmbH, Europaplatz 3, 69115 Heidelberg, Germany. If you have any concerns regarding our products, please contact ProductSafety@springernature.com

Printed and bound by CPI Group (UK) Ltd, Croydon, CR0 4YY
30/09/2026
02225332-0003